Medical Science Series

ACHIEVING QUALITY IN BRACHYTHERAPY

Bruce Thomadsen

Departments of Medical Physics and Human Oncology,
University of Wisconsin-Madison

Institute of Physics Publishing
Bristol and Philadelphia

British Library Cataloguing-in-Publication Data

A catalogue record for this book is available from the British Library.

ISBN 0 7503 0554 1

Library of Congress Cataloging-in-Publication Data are available

Series Editors:
C G Orton, Karmanos Cancer Institute and Wayne State University, Detroit, USA
J G Webster, University of Wisconsin-Madison, USA

Production Editor: Al Troyano
Production Control: Sarah Plenty and Jenny Troyano
Commissioning Editor: Kathryn Cantley and Ann Berne
Editorial Assistant: Victoria Le Billon
Cover Design: Jeremy Stephens
Marketing Executive: Colin Fenton

Published by Institute of Physics Publishing, wholly owned by The Institute of Physics, London

Institute of Physics Publishing, Dirac House, Temple Back, Bristol BS1 6BE, UK

US Office: Institute of Physics Publishing, The Public Ledger Building, Suite 1035, 150 South Independence Mall West, Philadelphia, PA 19106, USA

Typeset in LaTeX using the IOP Bookmaker Macros
Printed in the UK by J W Arrowsmith Ltd, Bristol

The *Medical Science Series* is the official book series of the International Federation for Medical and Biological Engineering (IFMBE) and the International Organization for Medical Physics (IOMP).

IFMBE

The IFMBE was established in 1959 to provide medical and biological engineering with an international presence. The Federation has a long history of encouraging and promoting international cooperation and collaboration in the use of technology for improving the health and life quality of man.

The IFMBE is an organization that is mostly an affiliation of national societies. Transnational organizations can also obtain membership. At present there are 42 national members, and one transnational member with a total membership in excess of 15 000. An observer category is provided to give personal status to groups or organizations considering formal affiliation.

Objectives

- To reflect the interests and initiatives of the affiliated organizations.
- To generate and disseminate information of interest to the medical and biological engineering community and international organizations.
- To provide an international forum for the exchange of ideas and concepts.
- To encourage and foster research and application of medical and biological engineering knowledge and techniques in support of life quality and cost-effective health care.
- To stimulate international cooperation and collaboration on medical and biological engineering matters.
- To encourage educational programmes which develop scientific and technical expertise in medical and biological engineering.

Activities

The IFMBE has published the journal *Medical and Biological Engineering and Computing* for over 34 years. A new journal *Cellular Engineering* was established in 1996 in order to stimulate this emerging field in biomedical engineering. In *IFMBE News* members are kept informed of the developments in the Federation. *Clinical Engineering Update* is a publication of our division of Clinical Engineering. The Federation also has a division for Technology Assessment in Health Care.

Every three years, the IFMBE holds a World Congress on Medical Physics and Biomedical Engineering, organized in cooperation with the IOMP and the IUPESM. In addition, annual, milestone, regional conferences are organized in different regions of the world, such as the Asia Pacific, Baltic, Mediterranean, African and South American regions.

The Administrative Council of the IFMBE meets once or twice a year and is the steering body for the IFMBE. The Council is subject to the rulings of the General Assembly which meets every three years.

For further information on the activities of the IFMBE, please contact Jos A E Spaan, Professor of Medical Physics, Academic Medical Centre, University of Amsterdam, PO Box 22660, Meibergdreef 9, 1105 AZ, Amsterdam, The Netherlands. Tel: 31 (0) 20 566 5200. Fax: 31 (0) 20 691 7233. E-mail: IFMBE@amc.uva.nl. WWW: http://vub.vub.ac.be/~ifmbe.

IOMP

The IOMP was founded in 1963. The membership includes 64 national societies, two international organizations and 12 000 individuals. Membership of IOMP consists of individual members of the Adhering National Organizations. Two other forms of membership are available, namely Affiliated Regional Organization and Corporate Members. The IOMP is administered by a Council, which consists of delegates from each of the Adhering National Organization; regular meetings of Council are held every three years at the International Conference on Medical Physics (ICMP). The Officers of the Council are the President, the Vice-President and the Secretary-General. IOMP committees include: developing countries, education and training; nominating; and publications.

Objectives

- To organize international cooperation in medical physics in all its aspects, especially in developing countries.
- To encourage and advise on the formation of national organizations of medical physics in those countries which lack such organizations.

Activities

Official publications of the IOMP are *Physiological Measurement*, *Physics in Medicine and Biology* and the *Medical Science Series*, all published by Institute of Physics Publishing. The IOMP publishes a bulletin *Medical Physics World* twice a year.

Two Council meetings and one General Assembly are held every three years at the ICMP. The most recent ICMPs were held in Kyoto, Japan (1991), Rio de Janeiro, Brazil (1994) and Nice, France (1997). The next conference is scheduled for Chicago, USA (2000). These conferences are normally held in collaboration with the IFMBE to form the World Congress on Medical Physics and Biomedical Engineering. The IOMP also sponsors occasional international conferences, workshops and courses.

For further information contact: Hans Svensson, PhD, DSc, Professor, Radiation Physics Department, University Hospital, 90185 Umeå, Sweden. Tel: (46) 90 785 3891. Fax: (46) 90 785 1588. E-mail: Hans.Svensson@radfys.umu.se.

This book is dedicated to Lucille DuSault, who taught me that the care of the patient always comes first, and Arthur Sweet, who taught me the value of quality.

CONTENTS

PREFACE

> Planning is important. Planning is important. The plan is useless, but the planning is essential.
>
> Dwight D Eisenhower

Is this a quality assurance manual or isn't it?

Certainly this book *is* a quality assurance manual. As such, it provides the reader with the steps for establishing a functional quality assurance programme in brachytherapy. However, those steps begin well before a check-off list of what to do for what type of case. The most important, and efficacious, steps address determining what the reader's particular facility really needs in terms of quality control and quality assurance—that is, creation of a quality management programme. No one has time and resources for a totally comprehensive quality control and assurance programme. Fulfilling the recommendations of some professional organizations would take sizeable increases in the staffing of many radiotherapy departments. Handling the job without abandoning other equally important tasks requires an analysis (and recognition) of a given facility's weaknesses, and building a programme to complement those areas, putting resources where they are most likely to prevent errors. Unfortunately, many regulatory bodies require practitioners to perform quality control procedures checking systems that have never failed, simply because they seem like important checks. The abstract and isolated list of checks may prevent some mistakes, but most erroneous treatments (that are noticed) happen in the face of a quality assurance programme. The problem usually is that the programme fails to address the realities of how the particular department functions. Related to the quote above, when things go astray (as they will), the plan, even the plan for what to do when things go wrong, usually is worthless. However, having gone through the planning gives an understanding of the situation that allows one insights for dealing impromptu with unexpected problems. Instead of simply giving lists of items to check (the reader can find those in the Task Group reports of the American Association of Physicists in Medicine listed in the references), this book suggests an approach to quality management, and provides discussion of the means and techniques for execution of a programme of quality control and assurance.

Most of the work presented in this text comes from other excellent persons devoted to achieving quality in brachytherapy. Notably, as seen from entries in the references, this author owes much to Jeffrey Williamson, Gary Ezzell and Eric Slessinger. Sankara I Ramaswamy spends his time working on quality instead of writing about it, and shared important pearls with me. In the compilation of this book, several companies provided figures that helped to clarify the sometimes rambling prose: Best Medical International, Medical Radiation Devices, Inc., Bill Kan, Mick RadioNuclear Instruments, Nucletron, Standard Imaging, 3M Company and Varian Associates. Much of my interest in the subject comes from interactions with co-workers who investigate cases with very serious shortages of quality: Judith Stitt, Barrett Caldwell, Rebecca McConley, Tonia Anderson, Partick Leammerich and Andrew Kapp. Bhudatt Paliwal lent encouragement and expertise during this writing. Of paramount essence in the production of this book was the support, most of all, of my wife, Dr Nancy Thomadsen. I would particularly like to thank my father-in-law, Arthur Sweet, the former Quality Control Supervisor with the State of Wisconsin, who inspired me to learn the difference between quality control and quality assurance, and their basic principles.

Bruce Thomadsen
Madison, WI, September 1999

CHAPTER 1

GENERAL CONSIDERATIONS IN QUALITY MANAGEMENT IN BRACHYTHERAPY

1.1. IMPORTANCE OF QUALITY ASSURANCE

Chances are that the reader, having picked up this book and gotten this far, needs no convincing that quality assurance, QA, serves an indispensable role in preventing patient injury and minimizing down time for the equipment used. Appreciation for the value of QA frequently follows some disaster that easily could have been avoided by a simple check beforehand. In many facilities, the quality assurance programme for brachytherapy seems driven by compliance with governmental regulations. Such a focus misses the opportunity to customize the programme to the individuality of the clinical practice, and may well fall short in important items not covered by laws and rules.

This text discusses programmes for assuring the quality of patient treatments in brachytherapy in considerable detail. Perhaps not all points discussed apply to all practices. The discerning readers will consider each point and decide the relevance in their own situation. Probably no institution will include all evaluations in this text in its routine procedures: doing so would simply consume too many resources and take too much time. The economics of the real world limits the efforts toward assuring quality and assuring that the treatment execution follows the therapeutic intentions. However, one mistake may cost a hospital millions of dollars in both legal fees and settlements, while a small fraction of this cost directed into effective quality assurance could avoid the expenses and detrimental publicity of such an event.

Values inherent in an effective quality assurance programme often evade monetary determination, but include peace of mind for the participants in the clinical procedure and mean that the treatments patients receive seldom deviate far from the ideal.

1.2. PRINCIPLES OF QUALITY ASSURANCE

Medical physicists generally have definite ideas about what quality assurance means. However, quality assurance constitutes a major field of study itself, outside the medical physics arena. Only of late have general quality considerations and the principles of quality assurance invaded medical practice, but they have done so in a major way, frequently applying quality control measures from industry in inappropriate manners. A familiarity with the more general conceptions regarding quality helps persons crafting a programme for application in their facility.

Terminology plays an important part in sorting out the various facets of this topic. Below follow some common terms:

Quality management

'All activities of the overall management function that determine the quality policy, objectives and responsibilities, and implement them by means such as quality planning, quality control, quality assurance, and quality improvement . . .' (ISO 1994). The goal of quality management is to achieve a desired level of quality.

Quality assurance

'. . . The activity of providing the evidence needed to establish confidence . . . that the quality function is being effectively performed' (Gryna 1988). Equivalently, 'quality assurance is: all the planned and systematic activities implemented within the quality system that can be demonstrated to provide confidence that a product or service will fulfil requirements for quality' (ASQC 1998). The goal of quality assurance is to demonstrate quality.

Quality control

'The operational techniques and activities used to fulfill requirements for quality' (Gryna 1988). Quality control (QC) consists of the tools used to meet the desired level of quality. QC follows the general process of (Juran 1988, p 2.9):

(1) Evaluating actual operating performance.
(2) Comparing actual performance to goals.
(3) Acting on the difference.

Much of what medical physicists call quality assurance falls more in the realm of quality control by these definitions. The American Society of Quality Control (1998) notes, '. . . often, however, "quality assurance" and "quality control" are used interchangeably, referring to actions performed to ensure the quality of a product, service, or process.' In fact, the Standards for Laboratory Accreditation of the College of American Pathologists (CAP 1987) include requirements for a 'Quality Assurance program to monitor and evaluate quality and appropriateness . . . [and]

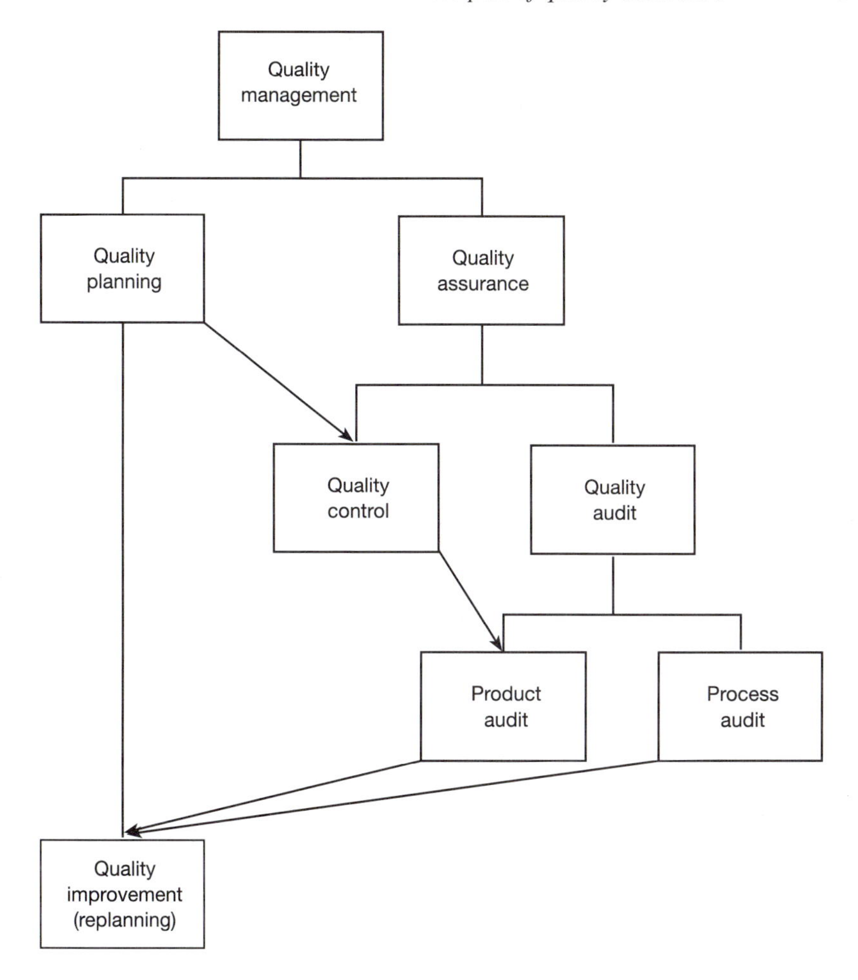

Figure 1.1. *A diagram of a sample quality management programme.*

identify and resolve problems,' and 'Quality Control that demonstrates the reliability of data'. Obviously, the two concepts share many features, and it often becomes unclear whether a particular action serves to control or demonstrate quality. To a great extent, the confusion obscures distinctions that clarify organizing processes. Since any discussion hinges on understanding the terminology, this text will try to follow the definitions above, where applicable.

A sample organization for quality management (QM) might look like figure 1.1. The first step, and one essential to the success of the process, consists of quality planning, setting out the methods and procedures to lead to high quality treatments.

Part of this planning process includes deciding how to define 'high quality treatments'. Defining 'quality' proves challenging at best. Contemplation of this question drives Phaedrus, a college instructor who wants to teach his students quality, to a nervous breakdown in *Zen and the Art of Motorcycle Maintenance* (Pirsig 1974). Phaedrus notes that, while quality evades definition, everyone has a sense for what things are high quality and what are low. Unfortunately, trying to establish procedures for ensuring quality without clearer ideas of a target goal could drive us crazy also. Juran, observing that quality has multiple meanings, suggests two as most applicable in this context:

(1) 'Quality consists of those product features which meet the needs of the customer . . .'.
(2) 'Quality consists of freedom from deficiencies' (Juran 1988, p 2.6).

He also observes that some define quality as conformance to standards or specifications. For brachytherapy, these lead us to defining quality as meeting the needs of the patient[1], remaining free from errors and satisfying the prescription, both as explicitly written and as implied.

Quality planning begins with decisions for each patient or type of patient regarding the needs for adequate and accurate treatment. The dose accuracy required varies with the tumour type and target site, and often with the treatment approach. These deliberations obviously require input from the physician involved with the treatment, but to a great extent rest with the medical physicist. Task Group 56 of the American Association of Physicists in Medicine (AAPM TG56 1997) recommends the following tolerances for brachytherapy delivery:

Positional accuracy: ±2 mm with respect to the appliance.

Temporal accuracy: ±2% for remote afterloaders (although all do much better, and manually loaded low dose rate cases have no problem in bettering this by a factor of 10).

Source-strength calibration accuracy: ±3%.

Dose calculation accuracy: ±2%.

Dose delivery accuracy: 5–10%.

The next step in quality planning entails developing the quality control procedures necessary to ensure both generation of a treatment plan that satisfies the patient needs, as reflected in the prescription, and accurate delivery of the plan. While the prescription serves as the primary documentation of the ideal objectives for the treatment plan, the physician probably desires more than actually written. Details of uniformity of the dose distribution, limits on doses to neighbouring structures

[1] While the medical physicist's customer in the business sense may be the physician, the view of quality management through this book focuses on the quality of the therapy delivered to the patient. The truly business aspects of quality management, as addressed in most of the quality management literature, will be left to other texts.

and dose gradients near the edges of the target volume often arise in discussion or follow some understanding derived through previous interactions between the physician and the medical physicist. Limitations on the whole treatment process, such as maximum times patients remain under anaesthesia, may play major roles in the treatment system. Part of quality planing includes establishing mechanisms by which all these parameters that delimit the treatment find accounting in the planning and delivery process.

Along the way through the treatment process, there should be tests and checks to assess that the process is proceeding properly and headed toward the correct outcome. Such tests include checks on the sources or equipment used for an application and evaluations of treatment plans. In part, these checks form quality control because they help direct the individual treatment to the successful outcome; however, they also become part of quality assurance since they demonstrate the correctness of the treatment.

An important and underutilized part of quality assurance is the quality audit. The quality audit consists of an independent review of the quality management programme. The independence of the persons performing the review allows recognition of weaknesses in procedures that the persons involved cannot see. If the facility will not bring in an outside expert to review the programme, an internal review provides some information on the quality of the brachytherapy programme . . . if the review looks with care at all aspects of the programme from the beginning to the end and compares each step to some generally accepted standard. Even with those provisos, an outside reviewer's perspective provides vastly superior recommendations (the inside team should have already implemented their ideas) and more credible support for establishing confidence in the quality of the programme (either by verifying the high quality of the programme or pointing to items needing improvement, or both). The auditors need the assistance and cooperation of the personnel at the facility, and the persons undergoing the audit should realize that, while receiving the critique may register as unpleasant, the audit facilitates improvement.

Aside from the external (even if conducted with internal personnel) quality audit as just described, the facility should also have an ongoing internal quality audit programme as part of quality assurance. This audit periodically reviews the quality control records of the treatments, looking for errors and trends indicative of potential for errors.

The quality audit contains two distinct parts: the process audit and the product audit. For brachytherapy the process audit entails analysis of the procedures followed through brachytherapy cases, covering all aspects of each type of treatment used. The auditor needs not only to be familiar with the customary approaches to the treatments, but also to understand the physics and clinical bases for the treatments to evaluate the justifications for, and ramifications of, deviations from customary procedures.

A product audit reviews a sampling of cases to assess whether the case achieved the objective specified. Such a review may involve an independent recalculation of doses, verification of treatment durations and inspection of prescriptions and inventory records. Depending on the activity of the facility under review, the review may look at all records or just a sampling. The aim of the product audit is to determine the probable 'failure rate', that is, the frequency with which the treatment differs from the intentions, and the seriousness of such discrepancies.

In figure 1.1, the plain lines connecting the boxes indicate the hierarchical relationships between the parts of the quality management programme. The arrows indicate where the output of a process becomes the input into a different process. As discussed above, part of quality planning entails determining the quality control procedures. Review of the quality control records forms an important part of the product audit. The results of the quality audits (both internal and external) provide the basis for quality improvement, i.e., quality re-planning to correct and amend the process to improve the quality of the treatments. Quality re-planning should include all players in the process: not that everyone's concerns can or should be met, but they should be considered.

The foregoing discussion has applied the concepts of quality management to brachytherapy. Of late, these concepts, as developed for industry, have found regulatory or accreditation organizations applying them almost directly to medical settings, frequently inappropriately. Several aspects differentiate the industry and medical situations.

- *Goal*
  The goal, sometimes referred to as the mission, of an industry is to make a profit; the goal in health care is to relieve the pain and suffering of the patient. This difference in viewpoint changes the interpretation of the quality concepts. Some of the applications of quality management in medicine have been generated through health-care administrators, and, unfortunately, retain much of the profit-oriented aspects associated with industrial settings.
- *Tolerance for error*
  In industry, as in medicine, quality comes with a cost. Figure 1.2 shows the relationship between the cost for a product and the assured quality for the product. In the figure, very poor quality comes with a high cost. In industry, these costs may correspond to replacement of parts in the field, liability and loss of customers. On the other side, extremely high quality also entails a high cost, from multiple inspections, production waste due to rejection or increased time in manufacture. The diagram implies that, for an industry, a minimum in the cost exists between the cost of failure and that of quality, and operating in that minimum maximizes profit. This approach considers a certain level of failure as acceptable. Brachytherapy procedures follow the same type of curve. Some level of error falls into the class of 'inconsequential', such as a 1 mm displacement in the dose distribution or a 1% discrepancy in the dose to a point. However, for the most part, medical procedures can accept very few failures, particularly as judged from the

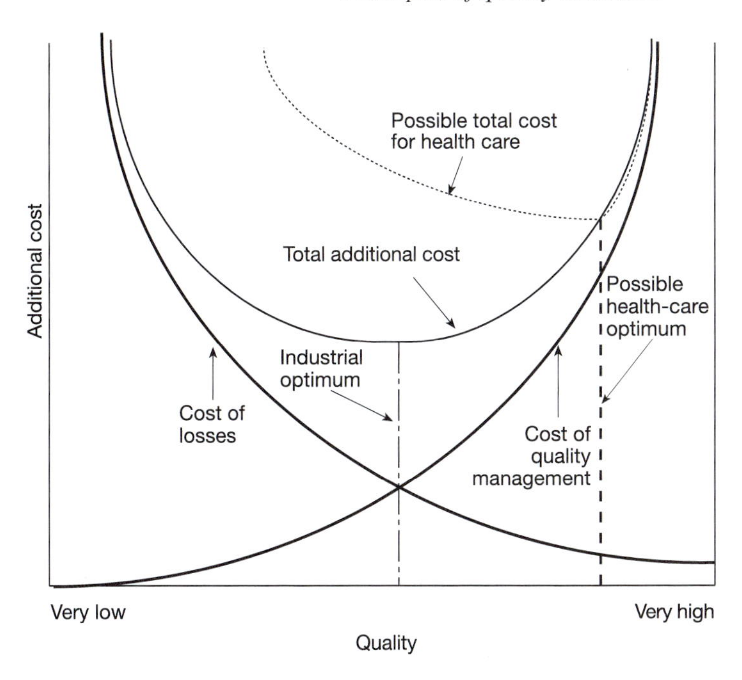

Figure 1.2. *The cost in addition to that just to deliver health care due to losses secondary to poor quality and to provide quality management. The chain line (— - —) shows the optimum balance between the two additional costs appropriate for industry, while the dotted line falls at a balance more appropriate for health care.*

perspective of the customer (patient). As a result, medicine operates far to the right on the curve, well above the minimum, and that, in part, accounts for the high cost of health care. Some analyses redraw the curve including litigation costs secondary to errors as the dotted line in the figure, arguing that medicine actually does operate in a minimum. This argument assumes that the only reason medical persons perform carefully is to avoid a lawsuit (or regulatory fines etc), ignoring that the parties involved enter the field to improve patient care. While no facility can afford to operate on the extreme right-hand side of the curve, medical operations find the balance differently than industry.

- *Separation of roles*

  Industries often consist of one team that produces the product, and a second team assigned to quality assurance. While this concept certainly plays an important role in medicine (as discussed under general quality control concepts), the same person often performs much of the quality control and quality assurance in medical settings.

Many innovations that work well in some industrial settings would prove disastrous in medicine. Take, for example, 'just-in-time' supplies, where supplies for an industrial operation come to a station just as they are needed. This minimizes the cost of storage and maintaining an inventory of parts, but in medicine a failure in the system could leave a patient without life-supporting care. Cross training may serve a function in industrial flexibility, but treating patients requires great expertise and experience.

1.3. ERROR

Any error can compromise the accuracy of treatment; an inaccuracy in treatment delivery results from an error. Errors generally fall into two categories: systematic or random.

1.3.1. Systematic error

Systematic errors stem from mistakes or errors in the general operation of a process. In brachytherapy, such errors include inappropriate algorithms in the treatment planning system, incorrect distances measured for localization radiography equipment or mistakes in establishing the calibration factors for the well chamber used for assaying sources. Errors of this sort affect all patients. Acceptance testing and commissioning should uncover these problems. Failure to ensure the accuracy of all the basic systems leads to general degradation of quality through all treatments.

1.3.2. Random error

Random errors include mistakes made during the process of an individual patient. Prevention of random errors follows the discussion below on error prevention. Almost all random errors stem from human errors.

1.4. ERROR REDUCTION

The quality assurance programme should employ both approaches to error reduction: error prevention and error interception.

1.4.1. Error prevention

For high dose rate brachytherapy (HDRB) treatment planning, error prevention amounts to reduction in the probability of errors in the input data entered into the treatment planning computer or treatment unit. The three techniques below help prevent errors from entering the data: *protocols, forms* and *a second person*.

1.4.1.1. Protocols. A protocol simply provides a formalized, standard set of expectations and procedures. When all persons involved follow a protocol, deviations from the protocol stand out as possible mistakes and items calling for investigation before proceeding to execution of the treatment. Some examples illustrate the variety protocols take.

Dose protocols

Doses prescribed for a given cancer often depend on the stage, grade or location of the disease. During treatment planning, the planner checks the prescribed dose and compares that dose to that listed in the protocol. For a cervical cancer treatment, for example, an application on a particular day may provide especially good separation between the uterus and the bladder and rectum. In response to such a good application, the physician might increase the dose for that day's fraction. However, such deviations from the expected value should lead the planner to question the prescription. The question not only entails whether the variant prescription accurately reflects the physician's intention, but how the physician determined the value, and how this prescription affects subsequent fractions. Unlike blindly fulfilling the prescription, the check on the correctness of the prescription provides protection for the physician against a mistaken prescribed dose.

Data-recording protocols

Endobronchial insertions using two catheters present the possibility that the identity of the catheters may at some time become confused or switched. As an example, faced with one catheter in the right lung and one in the left, there is no particular reason to expect the catheters to exit the patient from the nostril on the same side. The treatment for each catheter may fall at different locations along the catheter, with different length catheters, and possibly use different radial treatment distances or doses. To avoid confusion, the University of Wisconsin employs the following protocol for identifying the catheters:

(1) If the two catheters fall on different sides, the one on the patient's right uses the marker set indicating catheter 1, and the one on the patient's left uses the marker set indicating catheter 2.
(2) If the two catheters lead to bronchi on the same side, the more upper lobe catheter uses the marker set indicating catheter 1, and the more lower lobe catheter uses the marker set indicating catheter 2.
(3) If the two catheters fall on different sides, and the one on the left is in a more upper lobe than the one on the right, differentiation with respect to patient's side takes priority.
(4) The identity of the catheters is determined under fluoroscopy by inserting a marker train into only one catheter. The end of the identified catheter outside the patient is marked with tape and the appropriate catheter number. Then the correct marker trains are inserted into each catheter.
(5) A second person verifies the identity of the two catheters.

Variations from protocols

Not uncommonly, patients present with situations that fall outside normal expectations, and procedure specified in a protocol proves incompatible with the objectives of the treatment. Facilities must allow some freedom and flexibility to accommodate the realities of clinical operations. Even so, departures from protocols should follow a protocol on departures. Failure to provide guidance during variations and departures invites miscommunications, errors and injuries. For example, part of the protocol given above for identification for two-catheter endobronchial applications includes procedures to follow during deviations:

(1) If the expected identification routine proves inadequate, inappropriate or infeasible to apply, the person performing the identification must contact another person knowledgeable of the identification procedure to verify the validity of, and witness, the procedure used.
(2) The method of identification shall be written in the patient's chart.
(3) The identity of the catheters shall still be verified under fluoroscopy, and the catheters marked with tape at the patient's nostrils.
(4) The results of the identification shall be written in the patient's chart.
(5) The person performing the identification shall discuss the variance in the identification procedure and the results directly with the person performing the treatment plan.

1.4.1.2. Forms. The use of data forms serves several functions. The forms prevent lapses: omission of data due to momentarily forgetting what information the procedure requires. The guidance a form provides also helps prevent unintentional departures from protocols. By forcing the person taking the data to write information, the information, partially at least, is mentally processed. The act of writing makes the recorder more likely to challenge questionable data. Forms format information so all accessing persons know where to look and what to expect. Finally, forms can become part of a permanent record, facilitating reviews of the treatment in the future. Each type of treatment and often different parts of a treatment may benefit from a unique form. Using a form for procedures other than those that led to its creation frequently gives a false sense of security, believing that effective checks verify a treatment when in fact the forms may provide no control for the case in question. Frequent marginal notations on a form serve as one indication that a form conforms poorly in a given function.

The most indispensable form is the prescription. The description of the target site and dose distribution desired must be explicit enough to guide the dosimetry. While discussed in some detail in chapter 3, some information about the treatment becomes part of the prescription only by reference to the 'treatment plan'. The treatment plan, in this context, describes the brachytherapy application. The description should contain at least the following:

- A statement specifying whether the application is temporary or permanent, and if temporary, whether using high or low dose rate sources.
- The source material.
- A diagram of the implant, possibly showing different views as necessary to convey an understanding of its geometry. For multiple needle or catheter applications (including, for example, Heyman capsule packings), the diagram should label the needle tracks unambiguously. For treatments using remote afterloaders, the labels should relate to the treatment channel numbers.
- A description of the loading for each needle track or applicator part with respect to source strength and location. The description should be clear enough that any knowledgeable practitioner could duplicate the application exactly.
- The calculated dose distribution in as many planes as necessary to convey the shape of relevant isodose surfaces.
- The desired dose with specification of its relationship to the target (for example, as a minimal peripheral dose) and to the plotted dose distribution, and the time course of dose delivery.

This list does not apply to all applications. The process of customizing the list for a given treatment forms an important part of quality planning, and should precede the actual procedure.

1.4.1.3. Independent second person. An independent second person serves as one of the best error-prevention tools. A second person observing one who performs a function often sees mistakes the first does not, and allows immediate correction before the error propagates.

1.4.2. Error detection

Despite intensive efforts to prevent errors from entering into the planning process, some mistakes or misinterpretations are likely to slip through. To prevent these from causing injuries to patients (or regulatory hassles) takes a second line of defence to discover the presence of errors and correct them before treatment execution. The basic approaches to error detection use *comparison to standards and expectations, forms* and *independent reviewers*. Forms serve the same function as with error prevention, basically to assist the person performing the check to remember all the important steps. An independent reviewer, as with error prevention, provides an unbiased vantage for the evaluation. The designation 'independent' eliminates the second person involved in error prevention from performing the role as independent reviewer for error detection.

The basis for the evaluation of quality revolves around comparison of the item under consideration (the 'product' in the foregoing discussion) to some standards or expectations established *before* the evaluation began, and based on some information *external* to the item. For brachytherapy treatment plans, for example, the standard may be dosimetry tables from some well established system. Subsequent chapters discuss appropriate standards or expectations in considerable detail. As the prescription served as an important form for error prevention, it provides several of the expectations for error detection.

1.4.3. Guidance in the use of tools

Several concepts guide the development of the quality control tool.

Redundancy

Redundancy should not be equated with useless duplication. The duplication inherent in redundancy provides verification through matching the product of a second process to that of a first. The robustness of the verification strengthens as the two processes arrive at their answers by different paths. However, even if the two answers come from the same method, their agreement provides some check on random mistakes if the procedure at least produced the same result twice.

Forced attention

No technique forces an operator to pay attention. However, requiring interactions makes it less likely that the person performing a procedure mentally skips over important parts. Consider the design of forms used to guide a reviewer through an evaluation of a treatment plan. As a minimum, the reviewer needs to tick boxes indicating performance of the individual items in the review. Better yet, the reviewer should enter values into blanks on the form. Increasing still the attention required, the form could require the reviewer to enter the ideal value for a parameter, then enter the value from the plan, highlight the value on the plan and then enter the calculated difference between the two. Each entry forces the brain to partially process the value, and increases the probability of noticing a value out of range or unusual. Unfortunately, for frequently performed tasks, humans often slip into a mental mode of assuming things are correct and not paying the appropriate attention to warning signs.

Patient identification

While the prospect of treating the wrong patient seems remote to any facility that has never encountered such an event, this error happens not infrequently. In the United States, the US Nuclear Regulator Commission requires identification of the patient by *two* of the following methods:

- asking the patient his or her name;
- asking the patient for his or her address, and comparing that to the address in the patient's chart;

- asking the patient for his or her birthdate, and comparing that to the birthdate in the patient's chart;
- checking the patient's identification bracelet;
- comparison between the patient and a photograph of the patient;
- examining the name on the patient's medical record card.

Some of these methods require further qualification. In asking the patient's name, the patient should be asked, 'What is your name?' rather than, 'Are you Jane Doe?' Patients who either suffer dementia, who are hard of hearing, or simply preoccupied may, without really processing the question, reply in the affirmative to the latter. For the questions involving the patient's address or birthdate, or comparing the patient to a photograph, none of these methods count if the address, birthday or photograph originated at the same session as the check. These only serve as a check of the patient's identification if recorded previously, and the patient's identification checked positively at that time. Using the patient's medical record card for identification ignores the possibility that the cards may have been mis-filed. Not recognized by the US regulations, an additional method for patient identification includes affirmation on the part of one of the medical staff who knows the patient. All patients should have their photograph taken following identification for use in future identification, and to help prevent mistaken identity.

1.5. TRAINING

Training forms the first, and probably most important step in ensuring high quality brachytherapy treatments. Obviously, without proper training, the staff cannot execute their roles correctly. Indeed, the old saying, 'a little knowledge is a dangerous thing', applies well to brachytherapy. The records of misadministrations overflow with reports of persons performing functions for which they received training, but not enough. Insufficient training leaves trainees thinking wrongly that they know and understand the material, leading to inappropriate application of principles.

Training should begin with an explanation of the principles underlying a procedure, followed by instruction in the procedure with examples. After mastery of the procedure, the training should continue with examples of what can go wrong during the procedure, how to recognize when that happens, and how to recover. While none of the staff should be expected to substitute for persons of a different speciality, each member's training should include enough information on the other specialities' functions for an appreciation of the problems faced by all members of the team.

Simple lecture-type training sessions seldom provide the understanding necessary to execute a function adequately, particularly when the situation begins to deviate from normal. One effective model for training uses several parts to the instruction:

(1) *Lecture on background and principles*
This introductory section provides the information on the procedure necessary for understanding.

(2) *Demonstration*
Once the principles are understood, their application in the procedure can be demonstrated.

(3) *Examples by the trainee*
After the instructor demonstrates the procedure, the trainee performs the procedure with the guidance of the instructor.

(4) *Expanded examples*
Following several examples by the trainee, the instructor presents examples of greater difficulty with discussion of likely mistakes the trainee might make. As mentioned above, some of these examples should include recognition of, and recovery from, errors, and, if appropriate, steps to take in emergencies.

(5) *Examination*
All training should use examination to verify the trainee's understanding of the material taught. While the 'examination' may simply consist of the trainer's impression of the trainee's knowledge, a more formal process provides a better assessment. The author suggests examinations take two parts:

  (a) *Written examination*
  This mostly objective examination covers the main informational points about and steps of the procedure, common failure modes and rules governing the procedure. Passing grades should be established before administering the exam. Setting the passing mark at 100% reasonably reflects that to execute the procedure correctly requires understanding of *all* the points covered in the examination. However, the passing mark may vary with the procedure under consideration and the depth of the questions.

  (b) *Oral or practical examination*
  After passing the written examination, the trainer may delve deeper into assessing the understanding of the trainee through an oral examination, putting the trainee through the paces of the procedure, while the examiner throws complications into the case. The oral examination, where appropriate, would cover a run-through of emergency actions for various situations.

(6) *Supervised performance*
A trainee, after passing the examinations, begins a period of probationary performance of the procedure, that is, performing the procedure independently of guidance, but under the close supervision of the trainer. Initially, the supervision may entail watching every step of the trainee's actions, but, with successful completions of the procedure, the supervision diminishes, until the trainer declares the trainee 'graduated'.

In the terminology of this text, a graduated trainee becomes a *certified* practitioner of the procedure.

Emergency response forms an important part of training for many brachytherapy procedures, particularly those related to remote afterloaders, where problems with source control can deliver high doses of radiation to the patient or personnel. The training in these procedures should cover all of the likely potential problems, and discuss the situations that could lead to unlikely problems, and how to handle each. However, the training should stress that most probably, an emergency situation will *not* be any of the problems discussed, but something unthought of and new. Manufacturers usually address the problems operators conceive, but cannot solve the unknown. Still, practising actions responding to pretend problems gains experience in the nature of the device and the procedure that proves invaluable in the real, albeit different, situation. Again the Eisenhower quote rings true ('Planning is important. Planning is important. The plan is useless, but the planning is essential').

Periodic retraining serves as an opportunity to refresh procedures for persons who perform them infrequently, and to ensure conformality among those that perform the procedure. While regulatory bodies often specify the timing of retraining, the optimum time depends on the procedure, and how frequently the persons involved practise the procedure. For procedures performed routinely on a frequent basis, retraining or refresher training may not make sense at all, as long as there are periodic sessions at which the certified users discuss changes, unusual situations and points to keep in mind. For those procedures infrequently performed, refreshers or user meetings on an annual basis may recall important points forgotten. For any refresher, including users meeting for discussion, a written examination again provides a measure of the understanding achieved regarding the issues discussed.

Using the periodic refresher sessions to train staff in positions with high turnover (such as training nurses who care for patients containing radioactive materials) seldom provides adequately for their needs. Often by the time of the refresher, much of the staff would have had to deal with the situation without training. The refresher, or retraining, should be exactly that, and separate mechanisms used to train staff initially.

For a given procedure, or part of a procedure, a single member of the staff should be designated as the trainer to assure uniformity in training. The trainer must have been trained initially from another trainer, possibly at a facility that used the procedure earlier, and be actively involved in the procedure often enough to remain familiar with it. Retraining for the trainer comes from organizing and performing the training and refreshers for the rest of the staff, and, it is assumed, answering the questions that arise at these sessions.

The initial training for the trainer for a procedure, as noted above, probably came from attending training at a facility already practising the procedure. Such training should not be discounted or cut short. Aside from the tutorial in the procedure, participation in several clinical applications of the procedure should be included with the trainee performing the procedure under the watch of the trainer.

For some procedures, such as ultrasound-guided prostate implants, the training takes several days; training for high dose rate insertions may require a week or more of very intensive interactions. Training provided by manufacturers, while supplying a necessary foundation, seldom provides adequate depth for patient application without additional training at a clinical centre.

1.6. STAFFING

All radiotherapy comprises a team effort. Brachytherapy frequently requires a slightly larger team than external-beam treatments, and, unlike teletherapy, often all at the same time. The usual team consists of the following.

Radiation oncologists

As the physician involved in the case, the radiation oncologist defines the target and normal structures of concern, determines the type of procedure and applicator, inserts the applicator into the patient, assesses the adequacy of the application and prescribes the dose. Before, during and after the procedure, the radiation oncologist attends the medical needs of the patient, and evaluates the results of the treatment on follow-up.

Medical physicists

The medical physicist assists the radiation oncologist with defining the target and choosing the applicator; supervises localization and reconstruction imaging, the treatment planning dose calculations and optimization of treatment parameters and preparation of the source or sources; ensures the integrity of the applicator; supervises the loading of the applicator or delivery of the treatment and reviews the treatment for any errors.

Second, independent medical physicist

A second medical physicist reviews the treatment plan before execution to detect and intercept errors that entered during planning and calculation.

Dosimetrist

A dosimetrist may perform the treatment planning under the supervision of the medical physicist.

Nurse

A nurse may assist the radiation oncologist, and attend to the needs of the patient, during the application procedure. Nurses (usually different nurses) care for patients containing radioactive materials during radiation isolation in the hospital.

Radiographer

A radiographer images the implant site before the procedure for planning the treatment and/or after the insertion of the applicator for reconstruction.

Treatment-unit operator

For patients receiving treatments using a remote afterloader, an operator runs the unit under the supervision of the medical physicist. The operator may be a radiotherapist, a dosimetrist or the medical physicist. Nurses seldom have the background education and training to operate the unit safely in case of an emergency.

Radiation safety officer

The radiation safety officer supervises all the aspects of the brachytherapy programme dealing with regulatory issues and safety of personnel and the public. This includes source inventory, shipping and receiving. In many facilities the medical physicist performs this role.

Other physicians

Many procedures require the cooperation of physicians in other specialties, such as anaesthesiologists, urologists, gynaecologists or surgeons.

At some facilities, the medical physicist performs the functions above noted as 'supervising'. An earlier discussion considered the role of a second, independent physicist, and the chapter on high dose rate treatment planning looks at this role further. As this list shows, performing brachytherapy procedures properly requires a large staff. While some functions may be shifted to other members of the team, a facility unwilling to commit to supporting the personnel to execute the procedure without undue stress and operation outside the bounds of one's expertise should *not* offer or allow such treatments, but refer the patient elsewhere. Maintaining the training for the required staff exacts a considerable commitment on the part of the facility. Countries without national resources to support adequate brachytherapy teams should proceed with establishing brachytherapy programmes, but understand that quality is likely to suffer. In these cases, the patients probably benefit more from any brachytherapy than going without. The following approximate time commitments serve as a guideline for medical physicist staffing requirements:

- high dose rate (HDR) treatments—1.5 hours/fraction plus 0.5 hours/day for unit QA,
- low dose rate (LDR) implants—2 days/case;
  LDR intracavitary—0.5 days/case,
- Radiation Safety Officer (RSO) duties—0.5 days/case plus 0.5 days/month.

These times vary tremendously based on the types of case, the experience of the staff and the departmental organization determining who performs what functions. New procedures, of course, add considerably to the time.

1.7. ITEMS NOT COVERED

While this text addresses many of the important aspects of delivering high quality brachytherapy treatments, *all* aspects of brachytherapy enter into the final quality of therapy received by the patient. Important physical aspects *not* covered in this text include:

(i) brachytherapy application techniques,
(ii) source strength calibration theory and practice,
(iii) treatment planning,
(iv) localization and reconstruction procedures and
(v) dose calculation.

High dose rate brachytherapy entails many additional features not discussed herein. Many of the considerations relevant to HDRB will be familiar to users, but seemingly arcane to the uninitiated. The chapters in this book dealing with HDRB assume the reader understands the principles and operations involved with that modality.

Several texts discuss in considerable detail the aspects of brachytherapy omitted in this book, for example:

Williamson J F, Thomadsen B R and Nath R (ed) 1995 *Brachytherapy Physics* (Madison, WI: Medical Physics)
Williams J R and Thwaites D I 1993 *Radiotherapy Physics in Practice* (Oxford: Oxford University Press)
Nag S (ed) 1997 *Principles and Practice of Brachytherapy* (Armonk, NY: Futura)
Nag S (ed) 1994 *High Dose Rate Brachytherapy: A Textbook* (Armonk, NY: Futura)
Mould R F, Battermann J J, Martinez A A and Speiser B L (ed) 1994 *Brachytherapy from Radium to Optimization* (Veenendaal: Nucletron)
Godden T 1988 *Physical Aspects of Brachytherapy* (Bristol: Institute of Physics Publishing).

Many quality assurance programmes exist only to satisfy regulatory requirements. The discussions in this text essentially ignore regulatory considerations, firstly, because those regulations vary so markedly among jurisdictions, but mostly because a quality assurance programme should address the needs of the practice and properly address assuring the treatment delivered is correct and appropriate for the patient. Quality assurance mandated by regulations may provide these functions, but often does not, depending on the wisdom and experience of the person writing the particular regulations. Unfortunately, regulatory imposed QA procedures often serve little real function and sometimes inhibit good practice. This text also leaves consideration of general radiation protection and safety issues, such as shielding design and personnel monitoring, to the many good health physics textbooks.

1.8. TERMINOLOGY REVISITED

Much of the material published on quality assurance in radiotherapy follows the convention of the International Commission on Radiation Protection to use the term 'shall' as indicative of something one really has to do, and 'should' to indicate activities that are good to do, but not absolutely necessary. The discussion in this text uses should and must interchangeably as imperatives, and the term 'shall' has been generally avoided. As noted previously, all facilities *must* grapple with what to include in their quality management programme. Unfortunately, the absence of even the small steps, those that seem not very important, may rise to cause trouble in the wrong situation. To be effective, the quality management programme must stem from quality planning. Quality planning is not copying lists of procedure from any text (after all, the author is also subject to failures in human performance), but rather using the texts as guides in designing a programme specifically for the institution, addressing the local needs.

Brachytherapy covers a wide variety of procedures using many different approaches to position the sources near or in the target. Throughout this text, the various holders for the sources in the patient, (e.g., needles, catheters, intracavitary applicators) fall under the general term *appliance*. The term *needle track* refers to the path followed by an interstitial needle or catheter replacing a needle, and the loci of sources or source dwell positions therein. Along a needle track, or any afterloading appliance, *proximal* implies the end that the sources enter, and *distal* the opposite end. Since the dwell positions for a stepping-source remote afterloader function analogously to each source in conventional applications, general concepts including either situation usually will only mention 'sources' with the stepping-source dwell positions left understood. If the difference matters to a given discussion, the reference will be explicit.

1.9. SOURCES FOR INFORMATION

Some publications of particular interest in quality management in brachytherapy, and providing excellent overviews, include:

Williamson J F 1991 Practical quality assurance for low dose-rate brachytherapy *Quality Assurance in Radiotherapy Physics* ed G Starkschall and J Horton (Madison, WI: Medical Physics) pp 139–82

American Association of Physicists in Medicine: Task Group 40 (AAPM TG40) (Kutcher G, Coia L, Gillin M, Hanson W, Leibel S, Morton R, Palta J, Purdy J, Reistein L, Svensson G, Weller M and Wingfield L) 1994 Report 46: Comprehensive QA for radiation oncology *Med. Phys.* **21** 581–618

American Association of Physicists in Medicine: Task Group 56 (AAMP TG56) (Nath R, Anderson L L, Meli J A, Olch A J, Stitt J A and Williamson J F) 1997 Code of practice for brachytherapy physics *Med. Phys.* **24** 1557–98

American Association of Physicists in Medicine: Task Group 59 (AAPM TG59) (Kubo H D, Glasgow G P, Pethel T D, Thomadsen B R and Williamson J F) 1998 High dose-rate brachytherapy delivery *Med. Phys.* **25** 375–403.

1.10. HUMAN PERFORMANCE IMPLICATIONS FOR QUALITY MANAGEMENT

Just as quality management has recently entered the medical arena, so too has the field of human performance. Human-performance studies analyse incidents where something went wrong. Often untoward occurrences happen even with quality control procedures in place. The analysis attempts to find causes for the system failure, and develop new control procedures likely to prove more effective. While still in the preliminary phase, Thomadsen *et al* (1998) reported some conditions that commonly surround brachytherapy misadministrations (i.e., delivery of the wrong dose or the dose to the wrong site):

- design problems (frequently default settings);
- lack of information (probably due to training);
- expectation bias (operators failing to notice a change);
- distraction (due to pressures and other assignments).

Some more observations from studying the events indicate that:

- Many of the failure types are common in most industries.
- Errors usually are surrounded by complicating situations.
- Errors usually are surrounded by indicators, all ignored or rationalized by the principals.
- Errors frequently involve part-time staff.

Studies such as this should help direct limited resources into the most effective directions to avoid errors in brachytherapy.

1.11. SUMMARY

The process of planning an approach to maintaining quality in brachytherapy forms one of the most important parts of the quality management programme. During the planning for the programme the understanding of the criticality of each step develops. The quality management programme addresses the procedures to ensure quality (quality control) and methods to demonstrate that the quality for a procedure was upheld (quality assurance). Both arms of quality management assess treatments for errors. Error reduction makes use of techniques for error prevention (using protocols, forms and a second person) and error detection and interception (by comparison to standards and expectations, forms and independent reviewers). Training personnel in correct performance of tasks, of course, forms one of the most fundamental aspects of the quality management programme, but while necessary is not sufficient to prevent mistakes. Adequate staffing also forms a necessity.

CHAPTER 2

QUALITY MANAGEMENT FOR MANUAL LOADING, LOW DOSE RATE BRACHYTHERAPY SOURCES

One of the key ingredients for correct brachytherapy treatments is that the sources used in the treatment fulfil the expectations of the treatment plan. These expectations usually simply boil down to the source strength matching that in the plan, yet implicitly the plan makes assumptions about the source integrity and the construction of the source. All these aspects require verification.

For practical reasons, the extents of testing both necessary and practical to perform differ between long lived sources that make a permanent inventory and those short lived sources ordered for individual patients. Even with the differences, some tests remain the same, and discussion of these falls under the sources for a permanent inventory.

2.1. SOURCES FOR A PERMANENT INVENTORY

2.1.1. Acceptance tests

As with any radiotherapy equipment, new sources require acceptance tests to assure they match the expected specification. The tests include the following set.

2.1.1.1. Integrity (including leak tests). Source integrity includes inspection of the source for mechanical damage (i.e., damage that can be seen) and leak tests (looking for damage that cannot be seen). The inspection is fairly straightforward, checking the source for bends or dents. The welds must look smooth and in good condition. Most shipping containers for new sources protect the sources well from mechanical injury. Without significant damage to the outer container, damage to the source remains an unlikely situation.

Contaminated sources occur not uncommonly. Leak tests for new sources not only look for radioactive material escaping from the source through cracks or weak spots in the capsule, but also for extraneous radioactive material that may have contaminated the capsule during manufacture or during handling prior to shipping. Manufacturers test all sources for leaks prior to shipment. These tests usually use techniques called 'bubble tests' or 'immersion tests'. This procedure involves placement of the source into a container of hot liquid and watching for bubbles to form on the capsule surface as the air (or gas) in the source expands and leaks out. The bubble tests do not test for contamination on the surface of the source capsule.

Few hospitals have immersion test facilities, and usually test sources for integrity using wipe tests (sometimes called 'swipe tests'). For wipe tests, the surface of the source is rubbed with an absorbent material, and the absorbent material is then counted for radioactivity. Typical absorbent materials include filter paper, cotton-tipped swabs and specially made wipe test 'swipes', absorbent cotton pads, often with an adhesive backing attached to paper for recording information about the location or date of the test. Dry wipes may remove approximately 3% of radioactive material on a surface (Campbell *et al* 1993). Wetting the material increases the removal greatly, but with the provision that the polarization of the wetting solvent matches that of the material to be removed (Abelquist 1998). Alcohol works well as a solvent for both polar and nonpolar material because the opposite ends of the alcohol molecule exhibit the complementary properties. For polar contamination, water softening agents or detergents improve the removal fraction. Even with the appropriate solvent, the efficiency remains a function of the wipe material and the surface under investigation. Campbell *et al* found efficiencies of approximately 47% for stainless steel using water-saturated filter paper (Campbell *et al* 1993). Klein *et al* (1992) found that glass fibre wipe pads produced a higher removal efficiency than other wiping media. In the United States, regulatory bodies base the amount of measured, removable activity indicative of contamination, 0.005 μCi (185 Bq), on an assumed removal efficiency of 0.1. International standards often leave the determination of the efficiency to the user, setting the limits around the activity on the surface of a contaminated source (0.05 μCi, or 1850 Bq, by definition) rather than the measured activity on the wipe.

During the wiping of a set of sources, contamination can spread by using the same wipe for several sources. In addition, identifying the originally contaminated source becomes difficult if the wipe covered many sources. If the origin of the contamination is a leak, it may take several days for a measurable amount of surface contamination to build again and allow a second set of tests.

Significance of measured counts. Counting wipe tests often falls near the lower edge of the counting system's ability. The user must verify the ability of the counting system to actually detect an activity on a wipe test that indicates a problem. Based on the work of earlier investigators, Currie (1968) defined three conceptual quantities useful in discussing a counting system's sensitivity:

Critical level (L_C): the lowest reading on the system indicative of likely true activity in the counting sample.

Detection limit (L_D): the lowest activity in the sample that the system reliably detects.

Determination limit (L_Q): the lowest reading for which the system yields a reliable quantitative estimate of the activity in the sample.

Altshuler and Pasternack (1963) present a pictorial representation of the measured counts for samples with count rates near background measured in a hypothetical counting system. Figure 2.1(a) shows the frequency distribution for a series of measurements on two samples: the one on the left with no activity, and the one on the right with some small activity. The distribution of counts around an average value results from the random nature of counts in the background and that of the decay of the activity in the sample.

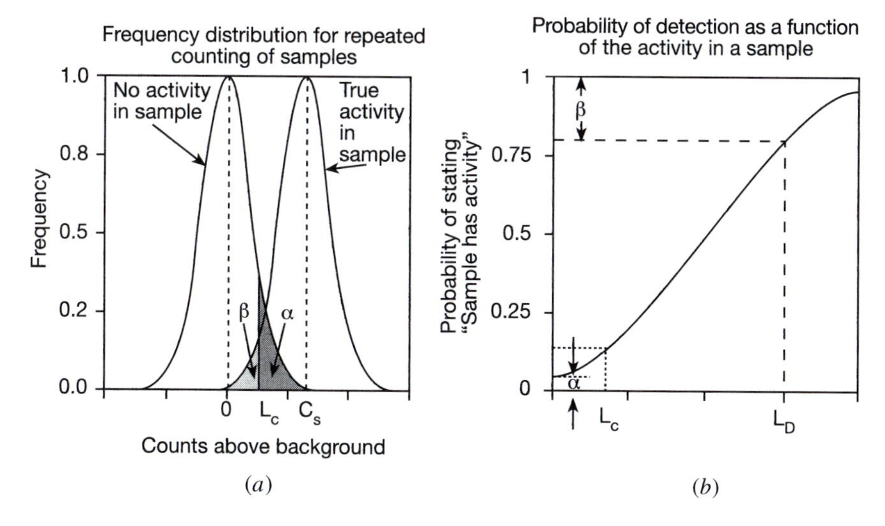

Figure 2.1. *Illustrations of the effects of counting statistics on detection: (a) distributions of counts resulting from multiple counts for samples with no activity (left) and true activity (right); (b) the probability of deciding a sample contains activity as a function of the true activity in the sample.*

A region of overlap exists between the two curves. A leak test presents the task of determining from a single measurement whether the sample contains an activity indicative of surface contamination on the source. Figure 2.1(b) shows the probability of implying that a sample actually contains activity based on a single measurement point. The abscissa lists the true activity contained in the sample, and the ordinate gives the probability of inferring a true activity greater than zero from the single count. Altshuler and Pasternack define two types of error one could make interpreting the measurement:

- type I error—inferring that the sample contains activity when it does not, and
- type II error—inferring that the sample contains no activity when it does.

The probability for the type I error, labelled α, simply equals the ordinate intercept where $A = 0$. The probability for a type II error, β, depends on a detection threshold set to delimit 'innocent' measurement from those indicating activity, *and* the true activity present. While the type II error depends on the detection limit, both errors depend on the relationship between measured counts and implied activity, and the system background count rate. Assuming that the frequency distribution of the count rate follows a standard normal distribution characteristic of radioactive decay and detectors counting background, the frequency of seeing a given count rate, r, with a standard deviation σ_r, equals

$$f(r/\sigma_r) = \frac{1}{\sqrt{2\pi}} \mathrm{e}^{(-0.5(r/\sigma_r)^2)}. \tag{2.1}$$

The probability that the count rate exceeds a value k_α is given by

$$P\{r/\sigma_r \geq k_\alpha\} = \int_{k_\alpha}^{\infty} f(r/\sigma_r)\,\mathrm{d}r = \alpha \tag{2.2}$$

and that the count rate remains less than a value k_β equals

$$P\{r/\sigma_r < k_\beta\} = \int_{-\infty}^{-k_\beta} f(r/\sigma_r)\,\mathrm{d}r = \beta. \tag{2.3}$$

For the normally distributed case, the relationships for these quantities follow table 2.1.

Table 2.1. *Values of k_α and k_β for probabilities of errors of type I and type II.*

α or β	k_α or k_β
0.005	2.576
0.010	2.326
0.025	1.960
0.050	1.645
0.100	1.282

An example illustrates how these quantities apply to the wipe test analysis. Consider the following data:

background count, $C_B = 700$
sample plus background count, $C_{B+S} = 728$
net counts for the sample, $C_S = 28$

The approximation of the standard deviation for these measurements is equal to

$$\sigma_B = \sqrt{C_B} = 26.5$$
$$\sigma_{B+S} = \sqrt{C_{B+S}} = 27.0$$
$$\sigma_S = \sqrt{C_B + C_{B+S}} = 37.8.$$

Finding what Altshuler and Pasternack call the minimum significant sample count (MSSC) and Currie defined as L_C comes from substituting the values into equation (2.2). The standard deviation in the denominator uses the expression for the background only since the MSSC considers the count boundary with samples containing no activity. That is equivalent to saying that the calculation determines the count rate corresponding to the cutoff for type I errors given the probability accepted for this type of error. The equation becomes

$$P_{S=0}\left\{\frac{C_S}{\sigma_B} \geq k_\alpha\right\} = \alpha.$$

Only the standard deviation for the background appears in the denominator since for the type I error the sample contains no activity, so the standard deviation of the sample equals that of the background rather than the sum in quadrature. The condition in the braces establishes the minimum count rate considered a significant indicator of activity in the sample,

$$C_S = \sigma_B k_\alpha = L_C.$$

From table 2.1, for an acceptable probability of inferring activity on the count with none actually present of 5%, $k_\alpha = 1.645$, and the MSSC $= 1.645 \times 26 = 43$. Since the net counts fall below L_C, this scenario indicates a clean source.

Possibly of more concern would be errors of type II, inferring no activity when the sample actually contains some. β gives the probability that with true activity L_D on the sample, the count rate falls below L_C, indicating no detectable activity. The probability for this condition follows

$$P_{S=L_D}\{C_S < L_C\} = \beta. \tag{2.4}$$

At the same time, equation (2.3) becomes

$$P_{S=L_D}\left\{\frac{C_S - L_D}{\sigma_{B+S}} < -k_\beta\right\}. \tag{2.5}$$

Solving equations (2.4) and (2.5) simultaneously for the critical value L_D, and assuming that $(k_\alpha + k_\beta)/\sigma_B \ll 1$ (see Altshuler and Pasternack for details),

$$L_D = L_C + \sigma_S k_\beta.$$

For the data in the example, the detection limit becomes $L_D = 43 + (37 \times 1.645) =$ 104 counts above background, or 804 counts.

While the National Committee on Radiation Protection and Measurement (NCRP 1978) recommends a value for α and β of 0.05, missing a leaking source 5% of the time seems a bit cavalier. Leaving the probability of mistaking a sound source as leaking as 5%, but reducing the miss rate to 0.5% in the above example changes the values to

$L_C = 43$ (since we did not change α), and
$L_D = 43 + (37 \times 2.576) = 138$ counts above background or 838 counts.

While it may seem strange that the counts increased when the desire was to make the system more sensitive. The correct interpretation is that the sensitivity of the counting system must produce this count differential between background and a sample with the limiting count rate in order to detect reliably the contamination at the threshold limit. Seen this way, increased sensitivity requiring an increased differential count rate makes sense. The problem then becomes achieving this differential, as addressed in the next section.

Counting times. Since the square root of the counts approximates the counts' standard deviation, and for most counting systems the only way to increase the count for a sample is to count for a longer time, the counting time becomes critical in establishing the sensitivity of the wipe test procedure. The previous section discussed determining the count level corresponding to significant indications of activity in the sample. The conversion from counts to activity at some point requires measurement of the signal from a calibrated radioactive source. Several vendors market sources with a calibration traceable to the National Institute of Standards and Technology, usually with units of activity (Bq), and suitably long half-lives to remain useful for long periods.

Going from the activity of the calibrated source to the conversion factor for the wipe test radionuclide requires accounting for the relative number of photons the counter is likely to record from each. Take, for example, an ^{192}Ir wipe test using a ^{137}Cs calibration standard with a NaI scintillation counter. With the window on the pulse height analyser set to count photons between 200 keV and 700 keV, each ^{137}Cs decay produces 0.98 photons that fall within the window (Bureau of Radiological Health 1970 *Radiological Health Handbook* (Washington, DC: US Government Printing Office)).

Each ^{192}Ir decay produces 2.17 photons within the energy window. Thus, for the same activity, the ^{192}Ir sample produces 2.21 times the count of the ^{137}Cs standard.

The activity on the ^{192}Ir sample becomes

$$A_{^{192}Ir} = C_{^{192}Ir} W \frac{_{t=0}A_{^{137}Cs}\ _tD_{^{137}Cs}}{C_{^{137}Cs}} \tag{2.6}$$

where

$A_{^{192}Ir}$ = the activity on the wipe sample,

$C_{^{192}Ir}$ = the net counts on the wipe sample in the given counting time,

W = the ratio of ^{137}Cs photons within the window to ^{192}Ir photons, per decay of each,

$C_{^{137}Cs}$ = the net counts on the standard in the given counting time,

$_{t=0}A_{^{137}Cs}$ = the activity of the standard on its calibration date,

$_tD_{^{137}Cs}$ = the decay correction for the activity of the standard at time t after its calibration.

The foregoing discussion assumed a constant efficiency for the interaction between the scintillation crystal and photons regardless of energy, an unlikely situation at best. Because of the energy dependence of the crystal, determining the response of the counting system to the radionuclide potentially present on the wipe best uses a calibrated source with a similar energy. For example, for wipes of ^{125}I sources, a ^{109}Ca calibration source provides photons fairly close to the 27 keV of ^{125}I. Most higher energy radionuclides use ^{137}Cs. If a facility only has a calibration standard available with an energy higher than that of the radionuclide on the wipe, assuming the same interaction efficiency is a conservative approach, as seen in equation (2.6). As the energy of the photons increases, more of the photons escape the crystal without interacting. The loss of photons for the standard reduces the count in the denominator, increasing the conversion factor $W\ _{t=0}A_{^{137}Cs}\ _tD_{^{137}Cs}/C_{^{137}Cs}$, resulting in a higher value for $A_{^{192}Ir}$. Rearranging equation (2.6) solves for the counts corresponding to the activity limit for a contaminated wipe, $A^*_{^{192}Ir}$,

$$C^*_{^{192}Ir} = A^*_{^{192}Ir} \frac{C_{^{137}Cs}}{W\ _{t=0}A_{^{137}Cs}\ _tD_{^{137}Cs}}.$$

Usually, the activity of the standard far exceeds the low level consideration of counts considered in the previous section but the counts for the sample fall close to background. (The significantly higher count rate might result in some loss in efficiency due to dead-time pileup, but that, as with the energy effect, results in higher calculated activities for the sample.) Because the significance of the standard remains beyond question, the count of the standard in equation (2.6) can be replaced with the count rate, $\dot{C}_{^{137}Cs}$, by including the counting time, T, in the denominator, giving

$$A^*_{^{192}Ir} = \frac{C^*_{^{192}Ir}}{T} W \frac{_{t=0}A_{^{137}Cs}\ _tD_{^{137}Cs}}{\dot{C}_{^{137}Cs}}. \tag{2.7}$$

Solving for the counting time gives

$$T = \frac{C^*_{^{192}Ir}}{A^*_{^{192}Ir}} W \frac{{}_{t=0}A_{^{137}Cs}\, {}_t D_{^{137}Cs}}{\dot{C}_{^{137}Cs}}. \tag{2.8}$$

In equation (2.8), the sample counts, $C_{^{192}Ir}$, must satisfy the requirements of significance from the previous section. In this example, the time must be long enough to accumulate a gross count for the sample plus background of 838.

Radium source testing. Historically, leak testing for radium sources used a very different approach. Working on the assumption that small cracks in the source capsule allow the radium daughter, radon gas, to escape, the source is placed in a stoppered tube for several days with a material that captures radon decay products. Examples of suitable materials include activated charcoal, molecular sieves or cotton smeared with a layer of petroleum jelly (often used to stopper the tube, but care must be taken to prevent coating the sources with the jelly while removing them from the tube). Any counts above background in the capture material indicates a leaking source. Counting the samples for either photons or alpha particles provides the same information.

Still another alternative places the source in a gas-flow proportional chamber set to reject photon pulses, and only record alpha interactions. While both the methods described will uncover leaking radium sources, they fail to detect any radioactive contamination on the surface of the source that does not produce a gaseous, radioactive daughter.

2.1.1.2. Linear uniformity. All dosimetry calculations assume a uniform distribution of the radioactive material along the active length of a source. The rare exceptions to this case involve either ‘Indian club’ needle-type sources with a greater proportion of the radioactive material toward the needle tip or planar applicators (such as ^{90}Sr ophthalmic applicators). Autoradiography serves as the common technique to examine the uniformity of the active material. For this test, a piece of Kodak Ready-Pak™ film is placed under a diagnostic x-ray unit, and the source is placed on the film. The source material creates an image on the film under the source in proportion to the strength of the material in the near vicinity. An adequate image is produced using a time of 4 min if using XV film or about 1 min for XTL film. During this time, the film is also exposed using the radiographic unit with a technique for a three-phase unit of approximately 60 kVp and about 25 mA s m^2. After processing, the radiograph shows the image of the source capsule while the autoradiograph indicates the location of the source material in the capsule. Figure 2.2 shows a sample of such an image.

2.1.1.3. Assay of source strength and identification. For this discussion, a differentiation will be made between the calibration and assay of a brachytherapy source. The calibration will refer to establishing the strength of a source based on

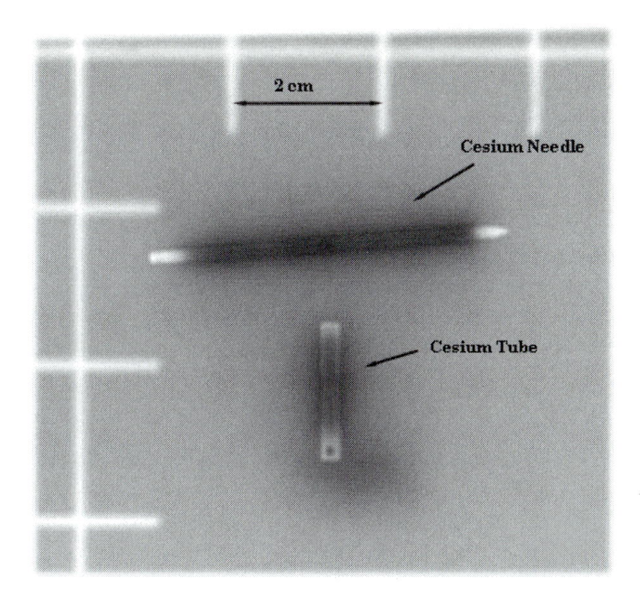

Figure 2.2. *Autoradiographs of caesium needle and tube sources.*

readings made using ionization chambers carrying their own calibrations in terms of charge collected per unit air kerma (or equivalent, such as exposure). Such primary calibrations require a great deal of time, care and expensive equipment, and usually occur only at standards laboratories. Even the standards laboratories perform such calibrations infrequently, opting for less time-consuming methods to provide information on a source's strength for users. Commonly, after performing a complete calibration on a source, the laboratory uses the now-calibrated source to establish a calibration factor for a well-type ionization chamber (figure 2.3). For subsequent sources *of the same model*, when placed *in the identical geometry*, the strength follows in direct proportion the net charge collected, as

$$S_{unknown} = Q_{unknown} \frac{S_{calibrated}}{Q_{calibrated}}$$

where

S indicates source strength,
Q indicates net charge collected after subtracting background,
unknown refers to the source being tested and
calibrated refers to the previously calibrated source used to calibrate the well chamber.

In this equation, the fraction represents the calibration factor, C_f, for the chamber.

For our purposes, a source strength established in this manner through comparison with a calibrated source, as $S_{unknown}$, defines 'assay', and forms a secondary

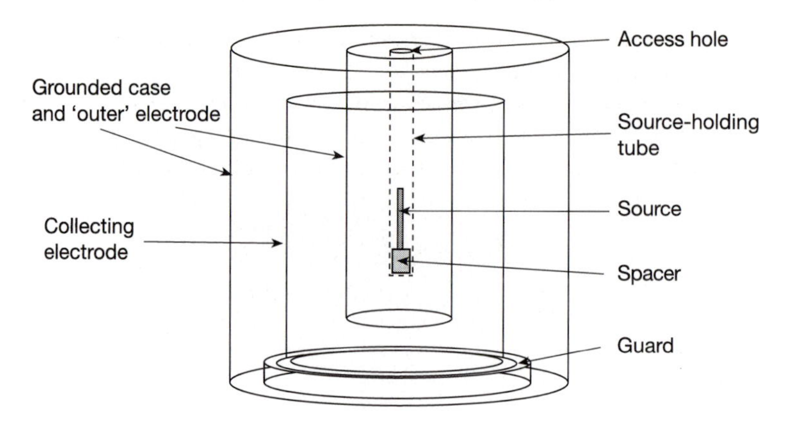

Figure 2.3. *Schematic drawing of a brachytherapy well chamber. Ions formed by the radiation are collected through most of the volume.*

calibration. An institution, on receiving a source with a secondary calibration, places the source in their own well-type chamber (often a nuclear medicine dose calibrator) to establish an 'in-house' calibration factor for use in assaying other sources of that design. These in-house assays become tertiary calibrations. As an alternative, an institution can send the dose calibrator to the standards laboratory to obtain the calibration factor. Such an approach, of course, requires doing without the device for the duration, usually about a week.

Obtaining an assayed source from, or sending the chamber for calibration to, a calibration laboratory costs enough that the practice remains an infrequent event. After establishing the calibration factor for the chamber, taking a reading with a long lived source in the well provides a consistency check for the instrument. For assays some time, t, after the initial determination of the calibration factor of the unit (at $t = 0$), the source strength follows the equation,

$$S_{unknown} = Q_{unknown} C_f \frac{Q_{t=0,consist} D}{Q_{t,consist}} \tag{2.9}$$

where $S_{unknown}$, $Q_{unknown}$ and C_f have meanings defined above, and $Q_{t,consist}$ indicates the reading for the consistency test source at either time t or $t = 0$. D in the equation corrects for the decay of the consistency check source from $t = 0$ to $t = t$, as

$$D = e^{-0.693t/T_{1/2}} \tag{2.10}$$

where $T_{1/2}$ = the half-life.

2.1.1.4. Well chamber quality management. The well chamber used for the assay of the source strength also requires a quality management programme. The evaluation below outlines the basic tests.

Spatial sensitivity. The chamber's response (charge collected per unit source strength) to a source depends on the source's position along the chamber's axis. Generally, the response rises as the source moves from the entry of the chamber toward the middle. At some point, the response plateaus for some portion of the chamber, and then decreases again as the source nears the bottom. Figure 2.4 shows a typical response curve.

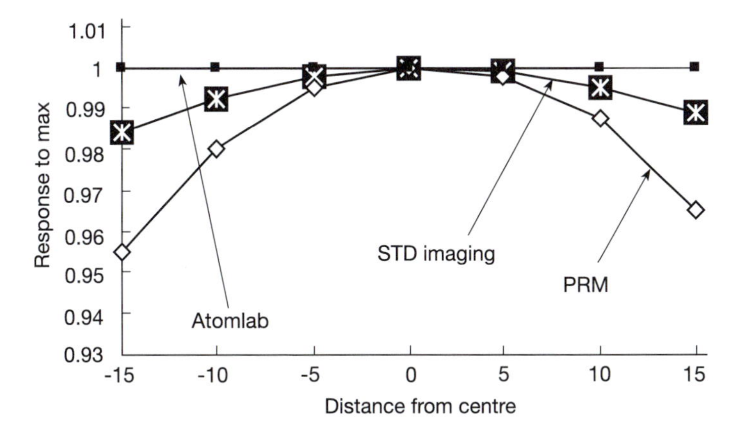

Figure 2.4. *Chamber sensitivity as a function of source position along the chamber axis (from DeWerd et al 1995). Figure used with the permission of the American Association of Physicists in Medicine.*

Locating sources in the middle of the plateau during calibration of the chamber and subsequent assays minimizes the uncertainty in the readings due to source positioning.

Collection efficiency. Most well chambers operate with collection efficiencies near unity, particularly for low dose rate sources. However, changes in the collection efficiency can indicate a problem with the chamber or operation (such as failing to turn on the bias battery). The collection efficiency should be checked at each use. One simple method uses readings taken at two voltages: one, R_o, at the operational voltage, and the other, R_h, at half that voltage. The correction for loss due to recombination becomes

$$C_{recombination} = \frac{3R_h}{4R_h - R_o}.$$

For a given range of readings, this correction factor should remain constant to within a couple of tenths of a per cent.

Stability with time. Most well chambers retain their calibration well over time with little drift. The chambers owe much of the stability to their very large collecting volume, where small changes in physical size translate into negligible shifts in calibration factors. However, some chambers have changed dramatically. Particularly, sealed chambers operated at pressures of several times atmospheric pose the risk of a leak changing the contained mass of gas and the resultant charge produced for a given source strength. For this reason, every use of the chamber should include a reading with a long lived check source. The reading from the check source, corrected for decay and atmospheric density if appropriate for the chamber, should match a reading performed at the time of calibration of the chamber. Deviation of more than 0.5% might indicate potential problems with the unit. As long as the check source readings remain within 2% of that at the time of chamber calibration, the assay of sources for clinical use may proceed, but the source of the change should be investigated.

Positional stability. Since the calibration factor for a given type of source depends on the source occupying the same location in the chamber as the source that established the calibration factor, positional stability becomes an important consideration. While the readings taken to check the stability with time should also detect a problem with positioning, measuring and recording the position of target position of a source in the well provides a quick check to differentiate between positional problems and other changes in response.

Energy response. Unlike a dose calibrator used in nuclear medicine, the well chamber for brachytherapy requires separate calibration factors for each source design. Thus, two types of source containing the same radionuclide often have different calibration factors. As a result, there is no energy response consideration.

2.1.1.5. Penetration and dose distribution. Most users assume that the dose distribution for a brachytherapy source follows that of published tables for the make and model of the source. Almost always the assumption proves valid. However, there have been cases of sources contaminated internally with different radionuclides. Such sources may produce aberrant dose distributions because of the different than expected photon spectrum. Unfortunately, testing for contamination proves extremely difficult. Two methods to investigate this include measuring the penetration of the radiation through absorbers such as lead sheets, or measuring the dose distribution in tissue-like material. Advantages of measuring the dose distribution include possibly verifying the anisotropy factors for the particular sources (which may differ from published data through variations in manufacture), and the fact that the dose distributions often are known very accurately. In addition, the quantity measured would be the actual quantity desired for treatment planning. Measuring the dose distribution, however, proves difficult and very time consuming. Verifying the energy of a source's emission by measuring its penetration through attenuators also presents several problems, the most serious of which is determining the most appropriate value for the attenuation coefficient for the measurement. Often, the values listed for shielding calculations assume extremely

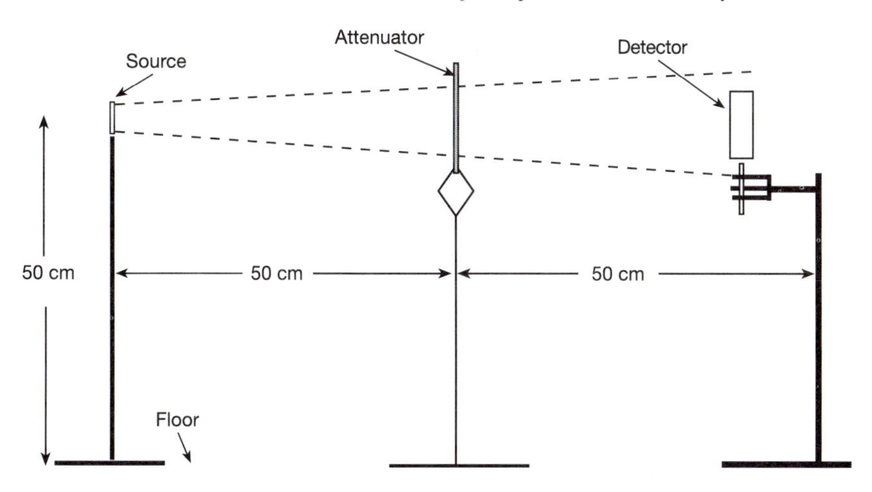

Figure 2.5. *Geometry for measurement of the penetration of the radiation from a brachytherapy source for use with equation (2.11).*

large sheets of attenuator (wall size) and err on the side of overestimating penetration. A good geometry for this test requires the attenuator fall far enough from the source to avoid appreciable scatter from the source, about 50 cm, and far enough from the detector such that little Compton scatter from the attenuator impinges on the detector (also about 50 cm, although for both distances 20 cm will do with little effect on the results). The shadow of the attenuator must cover the detector, of course, but should be little larger than necessary to minimize the scatter it generates. The entire experiment must remain far from other potential scatter sources in the room. Figure 2.5 illustrates the ideal geometry. The distances dictated by the constraints produce relatively small signals in the standard 0.6 cm^3 ionization chamber, necessitating the use of larger volumed chambers or survey meters (although Geiger counters are not appropriate for these measurements). For a reading R_0 without the attenuator in place, and R_t with an attenuator of thickness t between the source and detector, the linear attenuation coefficient becomes

$$\mu = -\frac{1}{t} \ln\left(\frac{R_t}{R_0}\right). \tag{2.11}$$

Table 2.2 provides some attenuation coefficient values for some common radionuclides for comparison. However, users should make some initial measurements for sources using their equipment, since that could change the measured attenuation coefficient.

2.1.2. *Periodic tests*

For long lived sources, some parameters should be tested periodically. Some regulatory bodies mandate periodic testing and dictate the period and the tests

Table 2.2. *Attenuation coefficients for some common radionuclides.*

Radionuclide	Attenuation coefficient (mm^{-1})
^{137}Cs	0.1155
^{192}Ir	0.231
^{125}I	27.7
^{103}Pd	86.62

required. Users or licensees must follow the regulated guidelines. However, regulations often aim to prevent large accidents and errors; 'good practice' may suggest more tests, more frequent tests or more stringent interpretations of tests. The suggestions below address the latter, and leave regulatory compliance to the user as a separate issue.

2.1.2.1. Integrity (including leak tests). The performance of this test follows the procedures above, but the actions to be taken following detection of activity on a wipe sample differ. A new source with a positive wipe test simply returns to the manufacturer. Indeed, a source identified as leaking during periodic testing also should be returned, in the manner discussed above. However, a manufacturer may not accept a source out of the warranty period. In some parts of the world, finding a disposal company or site becomes difficult to impossible. In such a case, the source first should be sealed in a nonporous container. A glass test tube makes a good, and often available, container. Plastic should be avoided since it may be more porous over periods of decades. Melting the end closed over a burner flame can seal the test tube. Great care must be taken not to heat the source, because it could explode, and personnel need to be shielded from the radiation from the source during the sealing. The sealed source should be placed in a shielded container, stored in the source storage room, and periodically tested with all the other sources. The source should *not* be buried or stored in an out of the way location where it could be forgotten and found by unknowing persons.

Upon obtaining a positive wipe test, the first step is to determine the radionuclide present in the wipe. A caesium source might show surface contaminated from an iridium source, and be misidentified as leaking. Most well counters used for measuring the wipes have either a pulse-height analyser or a multichannel analyser to determine the spectrum of the activity on the sample. If the spectrum matches that for the source's radionuclide, the source should be decontaminated by cleaning until wipe tests show no measurable activity above background, and then isolated in a test tube in a shielded container for about two weeks, and wipe tested again. A positive reading on the second wipe indicates a truly leaking source.

Following a negative second wipe, the source should stay in isolation for another three months and be tested again, with a positive result indicating a leak while a negative result clears the source of suspicion of a leak. At that time, all the other sources should be retested, since the activity on the original wipe test most likely came from one of them.

Leak tests should be performed six months after commissioning a new source, at its anniversary, and then annually. Some governments require more frequent testing.

2.1.2.2. Linear uniformity—first anniversary and then biennially. The construction of most modern brachytherapy sources renders shifts in the distribution of the radioactive material unlikely. Redistributions could happen due to manufacture defect, and could be most likely to occur in a short period. Redistribution also becomes more likely with sources stored with their axis vertical.

2.1.2.3. Assay of source strength and identification—first anniversary and then biennially. While checking the strength of a source is an easy time to also verify that the strength identification markings still show clearly, and that the source resides in the correct storage location. This becomes most important for needle-type sources with minimally different physical lengths (such as standard 'medium long' 5.2 cm and 'long' 5.7 cm needles).

2.1.2.4. Strength decay—first anniversary and then biennially. The measured source strengths should be compared with those predicted through correction for radioactive decay using equation (2.10). Deviations may indicate (1) presence of contaminating radionuclides in the source; (2) a change in the calibration of the well chamber or (3) an erroneous value used for the half-life in the decay formula. If checking the value used for the radionuclide's half-life uncovers no discrepancy, verification of the calibration of the ionization chamber becomes the next step. If there are different long lived radionuclides in the institution, these can serve as a cross check. Otherwise, a comparison of the institutional assayed values for short lived radionuclides compared to the manufacturer's values over the period of time in question might also show if the chamber's calibration changed. Intercomparison with another institution (not having any problems with source strength consistency) provides a reliable indication of the correctness of the ionization chamber's calibration. A discrepancy in chamber response determined through either analysis of short lived radionuclides or intercomparison with another chamber indicates that the well chamber requires recalibration.

After establishing confidence in the ionization chamber's calibration, if the sources still show deviant decay, the penetration of the radiation from the sources should be checked as described above. If the penetration changed compared with that measured during acceptance testing, the manufacturer should replace the sources, since the radionuclide purity most likely was not as per specifications. If the penetration remains the same as initially, new decay factors should be calculated and the sources' strength tested at 6 month intervals until developing a confidence in the effective half-life. That a source has a moderately different half-life than published values does not necessarily indicate a major problem in using the source, only that the strength requires more frequent verification. Substantial changes in the measured half-life, greater than 4% , do indicate significant radioisotopic impurities, and the manufacturer should replace the sources.

2.1.2.5 Inventory (just to see that the sources are in the correct location)—between assay checks. Many errors are discovered or prevented through entries in the source inventory. The inventory should contain three parts: one for periodic counting of the sources, one for recording source movement and one showing locations out of the storage room where sources find use. The first part helps assure that each source resides in its proper location (as well as accounting for all sources). For the counting, sources not in their storage location must be listed in the third part that shows each source in use at other locations. The second part lists an entry each time a source is removed or returned to the bank, indicating which type of source from which storage spot, who moved it, when and to or from where. Modern computerized databases simplify the process of sorting the movements and accounting for the sources, but the use of such a database should *not* eliminate the physical counting of the sources. Checking the inventory each time sources leave the bank helps assure that only desired sources go to a patient.

The frequency for the inventory depends on the use pattern. A very busy facility with several cases weekly would do well to account for all sources on a weekly basis. On the other hand, a department that only performs two or three procedures a year with the sources probably need only account for the sources with each case. As a general rule of thumb, the time between accountings should be about the same as that between every fourth patient, on an average, but no more frequently than weekly and no less than semiannually. Again, many countries regulate the frequency of accounting as a minimum.

2.2. SOURCES ORDERED FOR PARTICULAR PATIENTS

Sources ordered for particular patients pose different problems for quality management than those that become part of a permanent inventory. While the range of half-lives varies from 2.7 d for ^{198}Au to ^{192}Ir's 74 d, the general pattern of use finds the sources ordered shortly before a case and used soon thereafter. If removed from the patient or left over after the procedure, the user sends the sources back to the manufacturer. This usage pattern indicates that pre-application quality control measurements must be short, and the sources require no periodic testing.

2.2.1. *Integrity*

Typically the number of sources ordered for a patient falls between a score and a couple of hundred. Leak testing each source individually before use becomes a prohibitive commitment of time. Generally, a single wipe of the shipping container or vial serves to look for contamination. Iridium sources, were they wiped directly outside their plastic ribbon, might show removable contamination, since they are not closed (sealed) on the ends. While medical physicists seldom rely on manufacturers' tests or calibration, in this case, the pre-shipment leak test serves, with the institution's reception wipe test of the source container, as the test for source integrity.

2.2.2. *Linear uniformity—wire (not seeds, which are too short to matter)*

Linear uniformity becomes irrelevant for seed-type sources because they act much like point sources at distances greater than 1 cm. Iridium wire poses a real uniformity question. Bernard and Dutreix (1968) describe a device that looks at only part of an ^{192}Ir wire (either hairpin or interstitial wire for afterloading needles), and, based on calibration factors, yields the linear strength (strength per unit length) of the source (shown in figure 2.6). This device passes the wire through a channel in lead blocks that collimate the signal to a detector to a small portion of the wire. The system calibration depends on the length of the wire, due to two problems: penetration of the radiation through the collimation from the shielded portion of the wire and scatter from source along the wire reaching the detector. Calibration of the device assumes a calibration of a linear source with very uniform linear strength. Unfortunately, few standard laboratories calibrate ^{192}Ir wire sources, and the initial calibration of the device usually relies on a manufacturer's assay. The in-house assay of linear strength does provide a consistency comparison on subsequent sources as a check against a mistake by the manufacturer.

2.2.3. *Assay of source strength*

The procedure for the assay of the source strength depends strongly on the type of source.

2.2.3.1. Seed-type sources. Ideally, all sources would be assayed before use. Although not frequent, some users have found duds (non-active seeds) in a batch ready for implantation. Unfortunately, while practical for small orders, such as sources for an eye plaque, measurement of seeds in a well chamber consumes a large amount of time for large orders, such as sources for a prostate implant. The reports of Task Groups 40 (1994) and 56 (1997) of the American Association of Physicists in Medicine recommend sampling at least 10% of the sources, but not fewer than 10, as a measure of the conformality between the strength ordered and that delivered. For an order of 100 seeds, the sampling only gives a 1 in 10 chance of

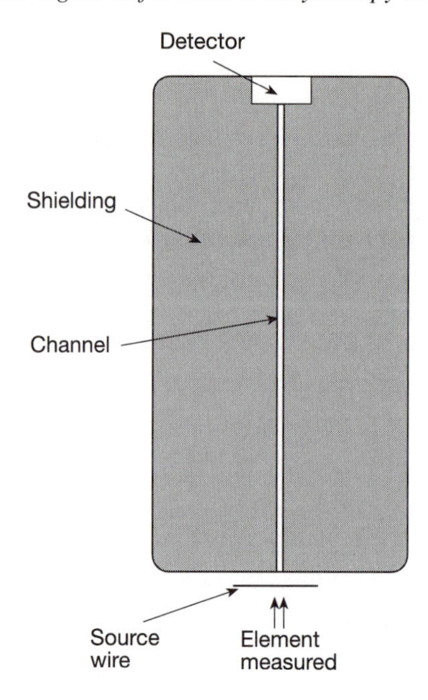

Figure 2.6. *The linear activimeter of Bernard and Dutreix (1968) for assessing the linear uniformity of iridium wire.*

detecting an aberrant source. While, for very large implants, the presence of a dud or two usually perturbs the target isodose surface little, locally the difference may be very large. Mellenberg and Kline (1995) describe a technique for measuring an entire batch of seeds at once to screen for significant errors.

The assay method described using equation (2.9) assumed that a standards laboratory provides a calibrated source, or calibrates a well chamber for a type of source, and based on that calibration rest all subsequent assays for sources of that type. Some sources in clinical use have no primary calibration at any standards laboratory. New brachytherapy sources often come into use before the standards laboratories establish reliable calibration factors and procedures for them. In-house assay procedures based on a manufacturer's calibration serve to assure consistency in source assay independent of the manufacturer, should the manufacturer's calibration procedures vary over time or contain unacceptably large uncertainties. Both situations have happened in the past and may well occur again in the future.

2.2.3.2. Seeds in ribbon or suture. Assaying sources in a plastic ribbon (as with most iridium seeds) or in resorbable suture (such as some ^{125}I seeds) present additional problems beyond those of assaying seeds alone. Attempts to focus a detector solely on one source in a ribbon, similarly to collimating the linear activimeter for

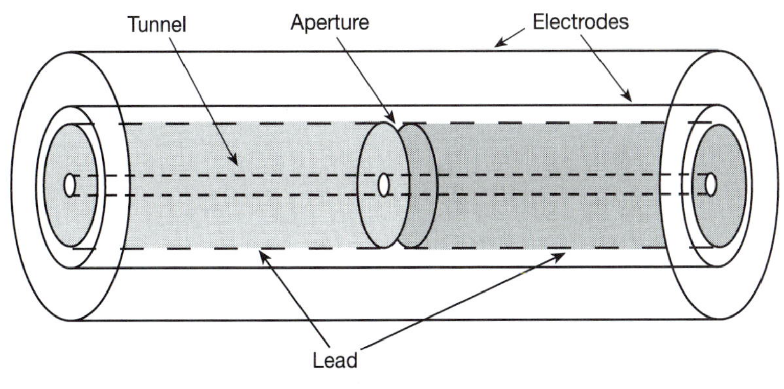

Figure 2.7. *Modified well chamber for assaying individual iridium sources.*

an iridium wire, fail because of the contributions of other sources even through lead shielding, or the poor sensitivity of a survey meter at the required distance. (Again, a Geiger counter would not be appropriate for this function.) Application of such a device for seed-type sources finds the system dependent on the number of seeds in the ribbon and the relative position of the seed along the train, as well as perturbations with differential loading. The system presents the additional problem of a lack of adequate signal strength at the detector (some of the sources may contain as little as 0.01 μGy m^2 h^{-1}—continuous wire usually comes with much greater strength per centimetre). Thomadsen *et al* (1997) describe using a shielded well chamber (figure 2.7), and applying a deconvolution/simultaneous-equation approach to sort out the strength of individual sources based on the response of a single, calibrated source passing through the chamber. While time consuming when performed manually, if automated this approach holds promise.

By characterizing the signal from a single source passed through the chamber (the deconvolution kernel), the signal for a source train is deconvolved to generate the positions of individual source signals (see figure 2.8). The procedure then solves n simultaneous equations

$$R_i = \sum_j^n s_j k_{j\to i}$$

where

n = the number of sources in the ribbon,

R_i = the reading when the ith source falls in the aperture,

s_j = the strength of the jth source (the unknowns),

$k_{j\to i}$ = the contribution to the signal when the ith source aligns with the aperture from a source in the jth position (from the deconvolution kernel).

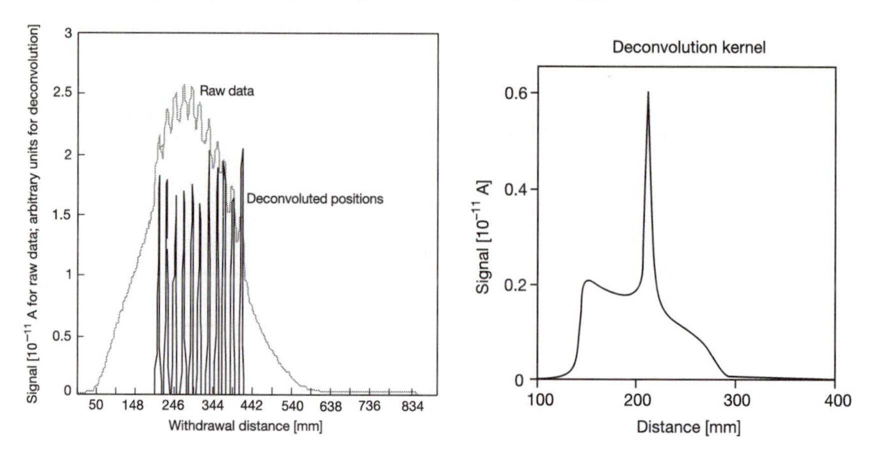

Figure 2.8. *(Left) The signal taken in the well chamber with the shielded insert as a ribbon of seeds pass through, and the seed positions established by deconvolution. (Right) The deconvolution kernel, established by passing one seed through the chamber.*

The information establishes a determined system with unique solutions. The authors report an accuracy for this procedure of approximately 3% for the assay of individual sources.

Seed sources in suture pose an additional challenge. The suture material around these sources may be either pliable for sewing through tumours or gauze-like applicators, or stiff, for insertion through implant needles. These products come presterilized using special processes because normal sterilization can degrade the suture. Thus, the source and suture must not make contact with any nonsterile surface.

Partially addressing the problem of calibrating these sources, one manufacturer (Standard Imaging, Middleton, WI, USA) makes a sterilizable shield that fits into a well chamber, allowing calibration of five of the ten seeds in a stiff suture together.

While not assessing the individual sources, this approach comes a little closer than looking at the whole set of ten, and provides some verification of the manufacturer's calibration.

2.2.4. *Penetration and dose distribution*

Because of the short time allowed between receipt and use of the sources, and the time consuming nature of testing for the radiation penetration and dose distribution, users rarely, if ever, investigate the correctness of these quantities.

2.3. SUMMARY

Table 2.3 summarizes the quality control tests directed toward sources used for low dose rate, manually loaded brachytherapy.

Table 2.3. *Quality control tests for low dose rate brachytherapy sources.*

Characteristic tested	Frequency of test	Method of testing
Sources from a permanent inventory		
Acceptance tests	*Performed at receipt*	
Integrity		Wipe test with alcohol-moistened filter paper
Linear uniformity		Autoradiograph
Assay of source strength		Calibrated well chamber
Identification		Visual inspection
Penetration		Attenuation measurement
Periodic tests		
Integrity	After 0.5 year, 1 year and then annually	Wipe test with alcohol-moistened filter paper
Linear uniformity	After 1 year then biennially	Autoradiograph
Assay of source strength and identification	After 1 year then biennially	Calibrated well chamber and visual inspection
Strength decay	After 1 year then biennially	Comparison of assay with calculation
Inventory	Approximately every 5 patients	Account for source locations
Sources ordered for particular patients		
	All tests at receipt	
Integrity		Wipe test of shipping container and manufacturer's certificate
Linear uniformity—wire		Measurement of small unshielded segments
Assay of source strength		
Seed-type sources		Well chamber (no less than 10% or 10)
Seeds in ribbon or suture		Not currently specified

CHAPTER 3

QUALITY MANAGEMENT FOR A TREATMENT PLAN

Evaluating a treatment plan before execution consists of assessing two basic aspects of the plan:

(1) the correctness of the plan and
(2) the appropriateness of the plan.

The first procedure prevents the execution of an error; while the second prevents delivery of a treatment one would rather not have used.

The principles discussed in this chapter apply to all brachytherapy approaches. However, this chapter focuses on low dose rate brachytherapy in particular. Treatment plans for high dose rate applications involve additional considerations, discussed in chapter 6.

3.1. EVALUATING WHETHER THE PLAN CONTAINS ERRORS

Errors enter treatment plans in various ways and with markedly different effects. Intercepting those errors after the completion of the plan requires comparing the data on the plan with some standards. In some of the checks the intended values serve as the standards, while in others the standards come from information external to the treatment plan. Each set of standards complements the other to provide closure to the evaluation.

3.1.1. Direct checks on the input data

The first checks simply review that the data used in developing the treatment plan correctly reflect the intended values. The basic data include the following.

3.1.1.1. Source strengths. Of course, for the simpler, uniformly loaded implants, checking the source strength usually requires only reviewing a single number on the

Table 3.1. *Conversions between conventional source strength units and air kerma strength (AAPM TG43 1995).*

Source type	Conversion factor
Any source in mg Ra eq	7.227 μGy m^2 h^{-1} (mg Ra eq)$^{-1}$
^{192}Ir, stainless steel encapsulation	4.60 μGy m^2 h^{-1} mCi^{-1}
^{192}Ir, Pt–Ir encapsulation	4.80 μGy m^2 h^{-1} mCi^{-1}
^{125}I seeds	1.45 μGy m^2 h^{-1} mCi^{-1}
^{103}Pd seeds	1.48 μGy m^2 h^{-1} mCi^{-1}
^{198}Au seeds	2.03 μGy m^2 h^{-1} mCi^{-1}

computer printout. However, a modern, low dose rate, temporary, brachytherapy treatment may use many different source strengths to optimize the dose distribution. While some computer programs allow on-screen display of the source strengths through a colour code, track by track, some still only provide a listing of sources, leaving the user to associate each source with its position along a track. Unfortunately, such systems usually are the same ones in which errors entering a variety of source strengths happen most easily. Moreover, while checking each source strength becomes tedious, errors in individual source strengths in differentially loaded implants formed the most common type of datum-entering mistake in our practice. Although small errors in a few sources distributed throughout a large implant seldom make appreciable differences in the dose distribution, were the errors to congregate in one region, differences could be serious. Of particular concern, though, are errors in input that change the strength for a whole set of sources. For example, setting all ^{192}Ir seeds on the inside of a gynaecological template implant as a bunch to 0.51 mg Ra eq instead of 0.15 mg Ra eq would produce a dose rate likely to lead to serious, possibly deadly, complications. On some treatment planning systems this simple typing error may prove difficult to notice.

Conversion of strengths between different units or quantities opens additional avenues for errors. Particularly at the time of writing, computer systems and source manufacturers are adapting to conversion from the conventional units for source strength (e.g., mCi or mg Ra eq) to the newer unit (air kerma strength). The conversion depends on the radionuclide and the source design. Table 3.1 gives values for some of the common sources currently available.

While one calculation performs the conversion, inverting the operation (for example, multiplying instead of dividing) forms an easy mistake, as does reading the value for the wrong source type.

With menu-driven selection of source types, picking the wrong type of source for a given radionuclide, or the wrong radionuclide altogether, increases in probability, and requires verification during the review. A more sure procedure has the operator select the source from a menu, and then type in the source type in a separate field.

Another cause of errors arises in the more sophisticated planning systems that allow entry of source strengths as of some date, and make correction for the strength on the date of the implant. Such systems often have default assumptions regarding the date of the procedure or calibration, such as being the date of the planning if no other date is entered. However, changes in these dates or inappropriate default conditions can lead to erroneous source strengths or dose rates, particularly if sources are used from a 'bank' of existing sources listed in the computer. Mistakenly entering the date in the wrong form, such as in American or European format when the computer expects the other, also is a common problem. Programs that use pull-down calendars avoid this potential error.

3.1.1.2. Appliance. At first thought, an error in the identity of the applicator may seem unlikely. However, some aspects of many applicators often remain poorly documented. For example, most intracavitary gynaecological insertions use plastic spacers between the source and the mucosa for that part of the treatment in the vagina, usually in the form of ovoids in the vaginal fornices or cylinders around a tandem. In either case, the diameter of the spacer varies to suit the patient's anatomy, with the largest diameter accommodated by the patient used in order to increase the dose deep compared to the surface. Unfortunately, while the radius of the spacers determines the dose distribution in the tissue, by themselves they produce no image on radiographs. Thus, verification that the physician inserted the size of spacers reported remains elusive. One solution involves placing small markers on the surface of the spacers (Podgorsak *et al* 1993). Figure 3.1 shows examples of steel marker seeds imbedded in ovoids and vaginal cylinders (in these cases for HDR cases). The markers indicate the spacer radius according to table 3.2. As an additional advantage, the markers provide points that the planner can enter from radiographs for dose calculations, and these calculated doses compared with that generated as part of the dose distribution for verification of both the dose and the location of the dose. Care must be taken entering such markers, however, since their centre lies below the surface of the spacer producing a tendency for the operator to enter the points too close to the source track. During digitization, the operator must concentrate on the outermost point of the markers on both views. Figure 3.1 also shows the images of very thin stainless steel wafers between the vaginal cylinders. Often the prescription specifies the dose location related to a given cylinder.

Computer treatment planning systems that store whole applicators, or dose distributions from applicators, while saving time during planning, allow the possibility of selecting the wrong file. Errors such as selecting a 15° tandem rather than a 30° prove difficult to notice by inspecting the dose distribution yet can deliver markedly different doses to the rectum, bladder or superior bowel.

Similar mistakes can occur selecting templates for implants, especially between two rectangular-patterned templates with hole separations differing by less

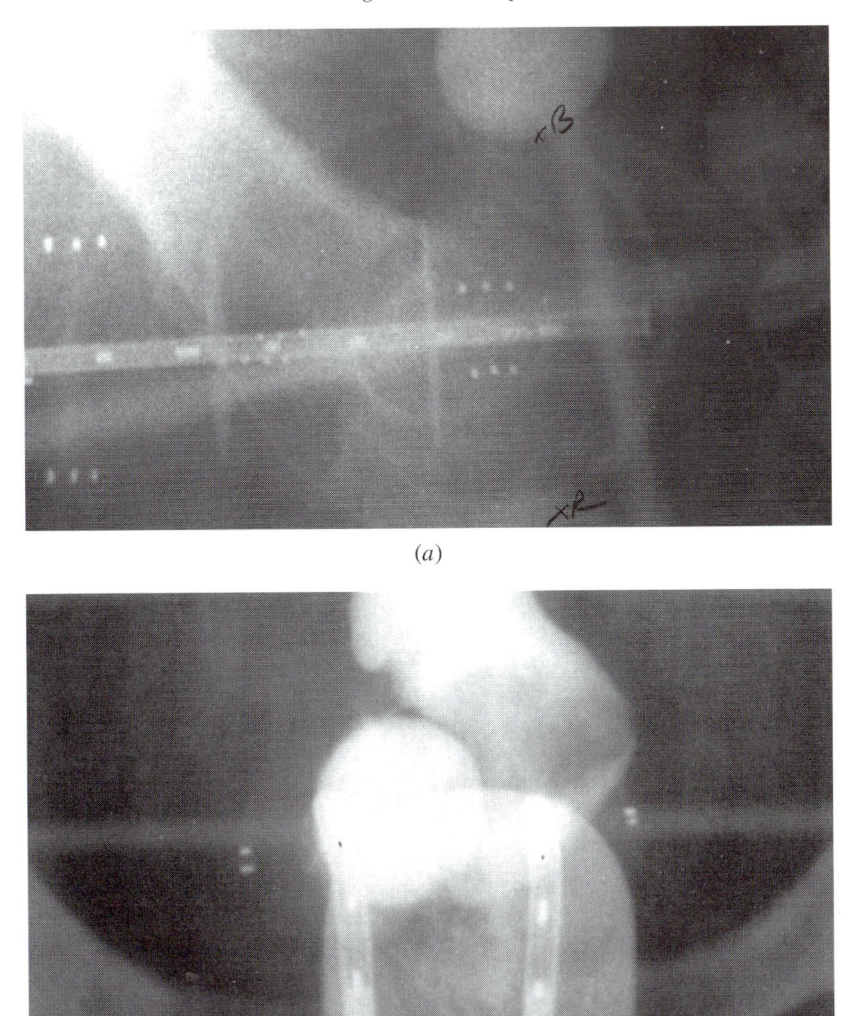

(*a*)

(*b*)

Figure 3.1. *Examples of markers inserted into plastic cylinders (a) and ovoid caps (b) to indicate the diameter of the appliance. The markers follow the coding in table 3.2.*

than 0.5 cm. In such cases, the treatment plan often uses the intended pattern, but the wrong template may be chosen in the operating room.

Table 3.2. *Coding for the size of indicators used with vaginal ovoids and cylinders.*

Number of markers visible	Diameter (radius) (mm)
1	20 (10.0)
2	25 (12.5)
3	30 (15.0)
4	35 (17.5)
5	40 (20.0)

3.1.1.3. Dose prescription and specification. While prescribing the dose for a given treatment generally falls in the realm of 'clinical judgement', that dose still requires verification by the person reviewing the treatment plan. A prudent physician would desire a second person assure that the prescribed dose contained no blunder. Presumably, the person performing the checks has extensive enough experience in the type of treatment under review to judge whether the prescribed dose falls within normal bounds. If that fails to be the case, reference texts provide instructive guidance. Many treatments follow clinical protocols, either departmental or interinstitutional, that indicate the prescribed dose based on the patient's disease and stage.

Part of verifying the prescribed dose includes checking that the dose used in the treatment plan follows the dose specification definitions intended by the physician writing the prescription. Many treatments follow systems, that is, sets of rules and guidelines (Thomadsen and Hendee 1999). A basic part of any system includes the method for specifying the prescribed dose, and different systems refer to the dose delivered in very different ways (that can differ by more than a factor of two in the amount of radiation the patient receives). Some examples of the method of dose specification from some of the more common systems include, for interstitial implants (see figure 3.2):

- *Manchester system*
  For a volume implant, the nominal dose (used for the prescription) is 10% higher than the dose at the periphery of the implanted volume (not counting corners), with the maximum dose between sources no more than 10% above the nominal. For a planar implant (one or two planes), the nominal dose is 10% higher than the dose at the periphery of a dosimetry plane 0.5 cm from the implanted plane (for a two-plane implant 0.5 cm in the direction of the opposite implanted plane), with the maximum dose on that dosimetry plane no more than 10% above the nominal.
- *Paris system*
  The reference dose (used for the prescription) equals 85% of the mean of the local dose minima between the sources (called the basal dose).

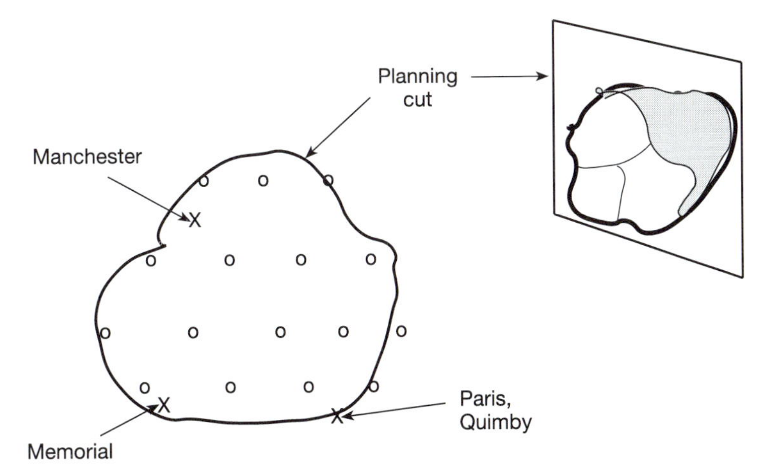

Figure 3.2. *Comparison of the locations for dose specification for some systems of interstitial implants. The plane on the left represents the cut indicated through the target on the inset to the right. The small circles indicate needle tracks.*

- *Quimby/Memorial system*
  The minimum peripheral dose is the dose just outside the periphery of the target volume for the Quimby system (Williamson 1998), and approximately at the outer needle line for the Memorial system.

For intracavitary insertions, the most common systems address cancer of the cervix explicitly:

- *Manchester system*
  Specifies a dose to the target location and the source loading pattern, as a function of stage of disease, as in tables 3.3 and 3.4.
- *M D Anderson system*
  Specifies the integrated disintegrations and the source loading pattern, as a function of stage of disease, as in tables 3.3 and 3.4.

In summary, evaluation of the prescribed dose addresses the questions

(1) Is the dose appropriate for the disease site and stage?
(2) Is the dose consistent with that used in similar patients?
(3) Is the dose specified correctly for any system in use?

A comparison with established systems similar to the application technique also allows verification of the total source strength developed in the treatment plan. Table 3.5 (from Thomadsen and Hendee 1999) gives values of R_V for volume implants from the Manchester and Quimby systems, with the former modified

Table 3.3. *Treatment specification as a function of stage for the Manchester and Anderson systems for treatment of cancer of the uterine cervix.*

M D Anderson system

Stage	Description	External beam (Gy): Whole pelvis	External beam (Gy): Boost	Max. duration (h): Tx No 1	Max. duration (h): Tx No 2	Max. mg Ra eq h	Max. vag. surf. dose (Gy)
Ia	≤3 mm, no vascular invasion	none	none	72	48	10 000	110
Ib–IIb	<4 cm	40[a]	to 60–62[b]	48	48	5600	120
Ib–IIb	small, good anatomy	20	10–20	72	48	7500	120
Ib–IIb	bulky	40[a]	to 60–62[b]	48	48	6500	140
III–IVA	good to fair regression	40[a]	to 60–62[b]	48	48	6500	140
III–IVA	poor regression	50	to 60–62[b]	48	24	5000	140

Manchester system

Stage	External beam (Gy)		Dose to point A (Gy): Tx No 1	Dose to point A (Gy): Tx No 2
I–II	none	none	38	38
III–IV	28.6	none	3500 mg h	38

[a] Dose delivered at 2 Gy/treatment, or alternatively 45 Gy at 1.8 Gy/treatment.
[b] Dose to pelvic wall including contributions from other external beam and intracavitary treatments.

to express the prescribed dose at the periphery, since most modern practitioners follow that practice. The R_V gives the strength of sources required to produce 1 Gy to the prescription point. This table applies to medium to high energy photon sources such as ^{192}Ir and ^{137}Cs. The source strength predicted by these tables follows

$$\text{total strength} = \text{dose rate} \times R_V .$$

Planar implants become more complicated in the specification of the dose for different systems. For Manchester-type implants (i.e., somewhat optimized), table 3.6 (Thomadsen and Hendee 1999) gives values for R_A, again modified for the dose prescribed to the periphery. R_A for planar implants performs an analogous

Table 3.4. *Source strength loading guide for the Anderson and Manchester systems for treatment of cancer of the uterine cervix.*

M D Anderson system		
Part of application	Size of part	Loading in mg Ra eq[a]
Intra-uterine tandem	Extra long	10–15–10–10 or Blank–15–10–10
	Long (normal)	15–10–10
	Short	15–10 or 15–15
Ovoids by diameter	Mini (0.8 cm radius on the lateral side)	5 or 10 each[b]
	2.0 cm	10 or 15 each[b]
	2.5 cm	15 or 20 each[b]
	3.0 cm	20 or 25 each[b]
Manchester system		
Part of application	Size of part	Loading in mg of Ra[a]
Intra-uterine tandem	Long	15–10–10
	Medium	15–10
	Short	20
Ovoids	Large	22.5 each
	Medium	20 each
	Small	17.5 each

[a] Loading for the tandem represents from left to right the cephalad-most source to the inferior, respectively.
[b] The loading for the ovoids depends on the vaginal surface dose rate necessary to keep the vaginal dose less than 140 Gy over the treatment.

role to R_V for volume implants, where the total strength is given by

$$\text{total strength} = \text{dose rate} \times R_A.$$

For volume implants using ^{125}I or ^{103}Pd, the total strength follows the guidelines in table 3.7 (modified from Thomadsen and Hendee 1999, based on Anderson 1976 and Anderson *et al* 1985, 1990, 1993). The total strength from the treatment plan and that calculated by one of the systems should correspond to within 5%. Differences of greater than 8% should initiate further evaluation of the generated plan.

3.1.1.4. Dose rate and treatment duration for temporary applications. The discussion of appropriateness of the dose rate selected for the delivery of the treatment follows in a later section. Only the correct transfer and use of the selected dose rate concern this discussion. Once chosen from isodose distributions, the dose rate

Table 3.5. *Values for R_V in U h Gy^{-1} modified from the Manchester and Quimby systems. The values in this table for both systems specify the dose to approximately the same position as the reference dose for the Paris system.*

Volume (cm^3)	Manchester R_V	Quimby R_V	Diameter of a sphere (cm)	Manchester R_V	Quimby R_V
5	94	156	1.0	21	31
10	149	251	1.5	47	77
15	195	307	2.0	83	139
20	237	347	2.5	130	217
30	310	428	3.0	188	303
40	376	493	3.5	256	370
60	492	600	4.0	334	449
80	596	699	4.5	422	528
100	692	807	5.0	521	620
125	803	907	6.0	751	844
150	907	1016	7.0	1022	1109
175	1005	1133			
200	1098	1226			
250	1275	1380			
300	1439	1485			

leads to the treatment duration from the equation

$$T = \frac{-T_{1/2}}{0.693} \ln \left[1 - \frac{0.693 D}{\dot{D} T_{1/2}} \right] \tag{3.1}$$

where

D = the prescribed dose,
$\dot{D}$ = the prescribed dose rate and
$T_{1/2}$ = the half-life of the radionuclide.

This equation simplifies for the case where the treatment time is long compared to the half-life of the radionuclide, becoming

$$T = D/\dot{D}. \tag{3.2}$$

The use of the simpler equation causes no more than a 3% error in the dose if the treatment time using the equation remains less than about 8.5% of the radionuclide's half-life. Table 3.8 lists the limits for such 'short' durations for several common radionuclides.

For most normal applications, the correction for the decay of the sources *in situ* affects only a second order change. Larger, and not uncommon errors related to treatment duration relate to major blunders in unit conversion. One of the

Table 3.6. *Values of R_A for modified Manchester and Quimby system planar implants ($U\ h\ Gy^{-1}$).*

Quimby system rectangles Treatment distances (cm)							Modified Manchester system Treatment distances (cm)									
3	2.5	2	1.5	1	0.5	Area (cm²)	0.5	1	1.5	2	2.5	3	3.5	4	4.5	5
						0	28.2	112	252	448	700	1008	1373	1793	2271	2802
859	567	390	229	109	35	1	64.0	161								
						2	91.3	201	353	563	814	1127	1502	1924	2396	2935
						3	113	233								
907	618	445	268	145	62	4	133	262	435	657	913	1228	1613	2041	2509	3053
						5	152	288								
						6	167	314	505	736	1004	1322	1715	2153	2615	3164
						7	181	338								
						8	194	362	564	805	1088	1412	1812	2255	2714	3269
1001	695	504	322	184	88	9	208	384								
						10	221	408	617	869	1163	1421	1902	2354	2813	3371
						11	234	429								
1053	769	558	362	212	110	12	246	452	668	932	1235	1575	1988	2451	2907	3467
						13	258	473								
						14	271	493	719	991	1305	1651	2072	2540	2999	3564
						15	284	514								
						16	297	533	766	1048	1374	1724	2149	2627	3088	3656
						17	309	551								
						18	322	570	813	1102	1436	1794	2225	2711	3173	3752
						19	334	587								
1096	785	583	397	253	152	20	346	604	857	1153	1495	1864	2302	2792	3259	3841
						22	370	635	904	1205	1554	1929	2374	2869	3342	3931
						24	393	666	949	1256	1612	1993	2446	2944	3426	4018
1199	886	668	466	313	192	25										
						26	416	694	994	1307	1665	2060	2514	3013	3506	4101
						28	439	722	1036	1354	1719	2122	2582	3084	3581	4186
						30	461	749	1075	1400	1771	2185	2653	3153	3656	4269
						32	483	775	1116	1447	1822	2241	2719	3220	3734	4351
						34	506	804	1154	1494	1875	2299	2784	3286	3811	4426
1317	1065	781	571	395	269	36	525	828	1194	1542	1928	2355	2846	3351	3884	4504
						38	547	856	1232	1587	1978	2412	2907	3415	3953	4579
						40	568	879	1267	1631	2026	2467	2966	3479	4024	4653
						42	588	906	1304	1676	2074	2520	3027	3542	4094	4727
						44	606	932	1336	1719	2124	2573	3084	3602	4165	4798
						46	617	956	1372	1761	2171	2625	3140	3662	4232	4872
						48	645	982	1402	1804	2216	2676	3196	3724	4299	4944
						50	664	1009	1433	1844	2261	2728	3253	3784	4362	5015
1522	1181	968	734	561	439	52	683	1034	1464	1887	2307	2778	3308	3841	4427	5085
						54	700	1059	1495	1927	2354	2827	3360	3900	4489	5155
						56	718	1085	1524	1969	2399	2876	3413	3959	4553	5224
						58	735	1108	1554	2012	2445	2925	3464	4018	4616	5292
						60	753	1136	1584	2053	2492	2975	3516	4075	4680	5358
						62	770	1158	1612	2092	2534	3025	3568	4133	4742	5424
						64	788	1186	1639	2129	2576	3072	3620	4187	4807	5489
						66	805	1209	1666	2167	2620	3116	3672	4241	4869	5555

Table 3.6. *(Continued)*

Quimby system rectangles Treatment distances (cm)							Modified Manchester system Treatment distances (cm)									
3	2.5	2	1.5	1	0.5	Area (cm^2)	0.5	1	1.5	2	2.5	3	3.5	4	4.5	5
						68	822	1236	1693	2205	2662	3164	3719	4295	4926	5619
						70	838	1261	1720	2241	2707	3211	3767	4348	4985	5680
						72	855	1287	1748	2279	2752	3258	3816	4402	5042	5741
						74	873	1313	1776	2312	2794	3305	3865	4456	5100	5802
						76	890	1338	1804	2345	2836	3352	3915	4511	5160	5861
						78	907	1361	1827	2379	2879	3398	3964	4562	5218	5920
						80	924	1387	1851	2412	2921	3444	4011	4614	5273	5979
						84	957	1435	1902	2476	3006	3535	4105	4721	5386	6095
						88	990	1480	1954	2540	3091	3624	4201	4827	5496	6213
1971	1554	1304	1053	858	735	92	1024	1526	2006	2604	3174	3713	4294	4929	5606	6327
						96	1056	1571	2059	2662	3256	3798	4385	5028	5713	6442
						100	1088	1615	2107	2721	3338	3879	4473	5127	5819	6549
						120	1231	1846	2364	2994						
						140	1378	2066	2625	3267						
						160	1514	2271	2876	3518						
						180	1644	2464	3119	3775						
						200	1771	2655	3352	4038						
						220	1891	2832	3582	4288						
						240	2007	3013	3808	4542						
						260	2125	3186	4038	4798						
						280	2233	3352	4266	5047						
						300	2349	3528	4482	5301						
						320	2468	3695	4693	5548						
						340	2578	3865	4896	5786						
						360	2686	4029	5109	6015						
						380	2794	4194	5301	6236						
						400	2900	4351	5499	6462						

Manchester two-plane separation factors	
Plane separation (cm)	Factor
1.5	1.24
2.0	1.4
2.5	1.5

most common occurs during the transition from the time in hours, from the appropriate equation above, to the actual day and time for source removal, particularly accounting for a.m. and p.m. and days of the week shortly after midnight.

3.1.2. *Does the plan make sense?*

Often the treatment planning system calculates the dose distributions from data locating the sources, but without relation to any target volume or structures. Particularly in these cases, evaluating the dose distributions must carefully assess whether the dimensions of the treated volume would match those of the target volume if the two coincided (as one would hope they will during execution). At various slices the shape and size must compare to the target at the same relative location, and the three major dimensions should be measured. Even using planning systems that display anatomic images superimposed with the dose distribution, this

Table 3.7. *Some guidelines for total strengths for the doses specified for some selected sources. The ^{125}I values are based on the 1999 standard from the US National Institute for Standards and Technology.*

Radionuclide	Dose (TG 43) (Gy)	Range of d_a (cm)	Strength (U)
^{125}I	139	≤3 cm	$5.7d_a$
		>3 cm	$1.52d_a^{2.2}$
^{103}Pd	120	≤3 cm	$23d_a$
		>3 cm	$4.14d_a^{2.56}$
^{198}Au	$D = \frac{1.344}{\sqrt{V}}$	All	$1.344d_a$

Table 3.8. *Limits for treatment durations for which the use of the constant dose rate approximation (equation (3.2)) produces less than the stated percentage error in total dose delivered.*

Radionuclide	Approximate half-life	Maximum time for error of less than	
		2%	3%
^{198}Au/^{90}Y	2.7 d	3.8 h	5.5 h
^{103}Pd	17 d	1 d	1.5 d
^{169}Yb	32 d	2 d	2.5 d
^{125}I	60 d	3.5 d	5 d
^{192}Ir	74 d	4.5 d	6 d
^{137}Cs/^{90}Sr[a]	30 y	1.8 y	2.5 y

[a] Information given for ^{137}Cs/^{90}Sr to show that the use of the approximation never causes appreciable errors in clinical applications.

check remains important. One such case occurred where the increment size between CT cuts changed in mid-study, but only the information in the header passed increment size to the treatment planning computer. Thus the planning computer assumed a uniform slice separation for all images. Because the slice separation changed from 1 cm to 0.5 cm before the implant site, the entire target became stretched to twice its length, although the two dimensions remained correct.

3.1.3. *Indirect checks on the input data*

In addition to verifying the input to the treatment plan, the output also requires verification. This process is not to check that the computer calculated the doses correctly based on the input data; that should have been checked during acceptance testing of the planning computer system. Instead, evaluation of the output seeks to uncover blunders during data input not easily verified. The manner of addressing this check depends on the nature of the application. The methods discussed here certainly only represent a small fraction of the possibilities.

3.1.3.1. Calculation of the dose rate to selected points. Independent calculation of the dose to points in the treatment volume and comparison to those calculated by the treatment plan provides some assurance that no major errors entered the calculational process, if, and only if, the following two criteria are met:

- *The source and calculational-point coordinates are entered independently.* While it is tempting to copy the coordinates from the treatment-planning computer, using these can only catch a computational mistake by the computer, the least likely source of errors. One of the most likely steps for error entry occurs during digitization of source or calculational-point coordinates.
- *The calculational points lie close to the sources.* When the calculational points fall distant from the sources, the dose depends mostly on the total strength of sources used with little ability to distinguish errors in source placement that greatly modify the dose distribution in the target volume.

As a caveat, the dose calculated to points taken too close to a source or sources becomes extremely sensitive to very small differences in digitization, and large discrepancies may signify no significant errors.

3.1.3.2. Calculation of the dose distribution using a second planning system. In principle, running the dose calculation through a second treatment planning system simply extends the check calculations at selected points to an extremely large number of points, and the same two criteria apply. The usual procedure entails running the calculation, and superimposing the two dose distributions. Exact matches seldom happen because of slight differences in the calculational algorithms, stored data tables, plotting utilities or plane selected for calculation. Usually, because of the large gradients in doses present in most brachytherapy applications, evaluations of the agreement between the two plans amounts to measuring the differences in positions of given isodose-rate lines. One millimetre shifts represent normal variations, but discrepancies of 3 mm require investigation. In regions with little gradient, such as the middle of an optimized volume implant, differences in dose rate serve as the measures of agreement, with 2% often considered the allowed tolerance.

3.2. EVALUATING THE APPROPRIATENESS OF THE PLAN

Aside from looking for untoward errors that might have entered the treatment plan, a quality assurance review also should render a judgement regarding appropriateness of the generated plan for the treatment. The discussion below first focuses on the areas of concern in making such an evaluation, and then on the tools to help in the assessment.

3.2.1. *Concerns in evaluating the appropriateness of a treatment plan*

Three major concerns enter the evaluation of appropriateness of a treatment plan: coverage, tolerance and uniformity. All three intertwine, each affecting the others.

3.2.1.1. Coverage. An application that fails to adequately cover the target serves the patient poorly. Some implants require physical palpation by the performing physician during the implantation to delineate the target, most often in the head and neck region. Lack of contrast on CT or MRI may prevent planning before the procedure in these cases, and the catheters implanted become the indicators of the target volume. Except for cases of such an occult nature, the target determination should form one of the earliest parts of the treatment planning process, and the adequacy of target coverage for an executed implant judged against this pre-established volume.

Two general approaches to covering a target volume find widespread application. One approach, following the lead of the Manchester system, assures that the needle tracks fully enclose the target volume. Although following the source distribution guidance of the Manchester system, most modern practitioners specify the dose at the periphery of the target volume (i.e., the limits of the implant proper) rather than 10% higher than at the periphery as actually specified in the system. This approach assures coverage of the target, but at the expense of increasing the amount of normal tissues included in the treatment volume or raised to high doses.

The other general approach follows the principle of the Paris system, specifying the dose to an isodose surface that extends *outside* the implanted volume. Many anatomical sites lend themselves to such an approach, for example permanent implants of the prostate. Radioactive seeds placed outside the prostate risk loss or severe migration.

Both systems recognize that the dose rate along uniform-strength needle tracks falls below the treatment dose well before the end of the track. The Manchester system includes provisions for placing additional tracks at the ends and perpendicular to the main tracks to bring the dose rate back to that in the middle. Failure to be able to provide these additional crossing tracks requires extending the main tracks beyond the target to assure the end of the target comes before the fall in the dose. Most volume implants need about an 8% increase in the track length per uncrossed end; for planar implants, the increase runs about 11%. The Paris system routinely uses no crossing tracks, and suggests that the track lengths equal 1.4 times the treatment length. Optimized treatments, for example, ^{192}Ir seeds in ribbons with higher strength seeds in the end positions, may project the treatment dose adequately toward the ends of the implant to allow the sources to remain within the target volume.

Williamson (1991) calls attention to the relationship between coverage length for linear sources and that for seed-type sources in a row originally noted by Shalek and Stoval (1969) (see figure 3.3). Looking at a seed in the middle of the source train, each seed 'covers' a length equal to the seed separations, with

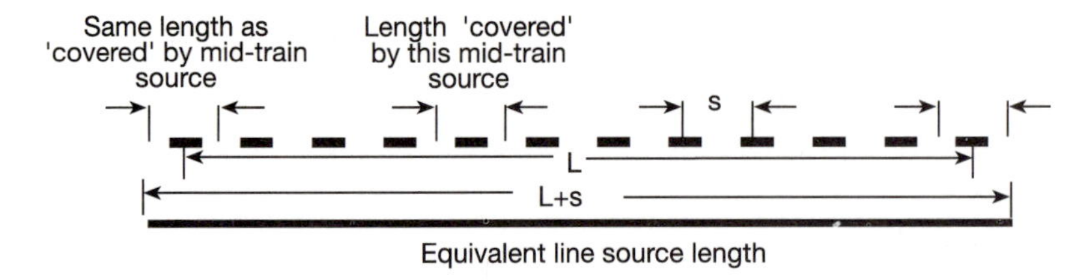

Figure 3.3. *The equivalence in length between a line source and a train of equally weighted seed-type sources.*

half that length on either side. Assigning this coverage length to the end seeds, the effective length of the seed train becomes the length from the centre of the first seed to the centre of the last seed plus the separation between the seeds. In the figure this effective length is equal to $L + s$.

Coverage in intracavitary treatments assumes a somewhat different guise. With all the sources essentially in the middle of the target, the dose falls continually with radial distance from the sources. Without an isodose plateau, no dose rate represents the treatment better than any other. Adequate coverage may simply require selecting a dose rate low enough to cover the entire target. Such an approach pays its price when considering the next concern, tolerance.

3.2.1.2. Tolerance. The main hurdle to curing cancers with radiation is the limiting tolerance to radiation of the surrounding normal structures. That radiation treats tumours *in situ* appears as an advantage compared to surgery that often leaves deforming tissue defects. Yet, that same fact, that the normal tissues must remain healthy to allow healing of the tumour site as the cancer regresses, means that radiation oncologists take the first principle of medicine very seriously: above all else, do no harm[1]. Figure 3.4 shows an idealized graph relating effects to radiation dose. In figure 3.4(a), the curve to the left represents the probability of tumour cure. The probability increases with dose, approaching totality at extremely high dose values. The curve to the right depicts the probability of some complication in the normal tissues. The two curves follow the same basic shape, similar to the integral of a normal distribution, although the finer details tend to differ. The dose indicated by vertical line 1 produces a cure in approximately 90% of the tumours, but at the cost of serious complications in 40% of the patients, an unacceptably large number. Reducing the dose to vertical line 2 brings the complication probability to about 5% (still possibly high depending on the nature of the complication), but yields only a 20% probability of cure. Successful treatment (a cure without complication) in this setting remains a rare event. Figure 3.4(b) shows some approaches to improve the success rate. Arrow 1 indicates a movement of

[1] Of course, some argue that the greatest harm for patients often is the continued growth of their tumours.

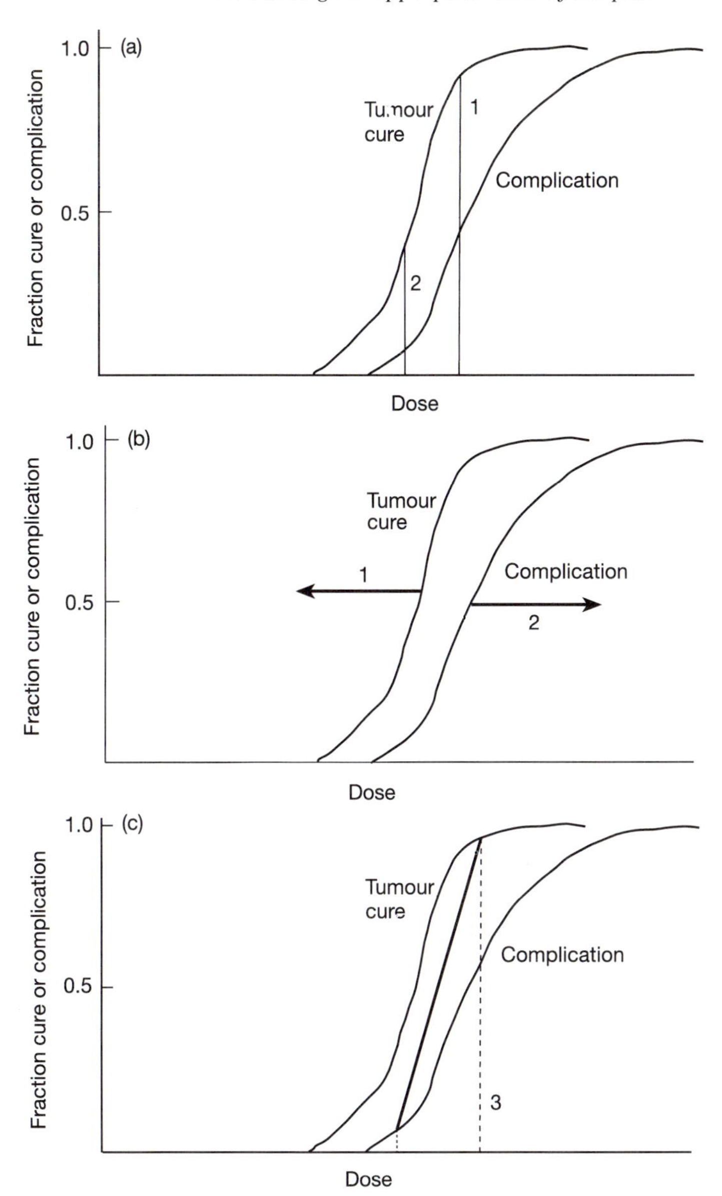

Figure 3.4. *Schematic illustrations of the response of tumour and normal structures to radiation doses. Use with the discussion in the text.*

the tumour response curve to the left, possibly through the use of tumour sensitizing chemotherapy agents or earlier detection and treatment. Such an action increases the separation of the curves, permitting doses with high probabilities of success. Alternatively, arrow 2 depicts the movement of the normal tissue curve to the right, maybe through the use of a radioprotective agent. The separation of the tumour and normal-tissue curves this way likewise improves the likelihood of successful treatment. Unfortunately, sensitizing and protective agents that work effectively and reproducibly remain few and limited in application. Figure 3.4(c) shows a different and time-tested approach to the problem. Success can be had by delivering a dose with a high probability of cure to the tumour (line 3) and a *different* dose to the normal tissues that has a low probability of complications. Of course, the production of the differing doses remains the main challenge of radiotherapy physics. Brachytherapy addresses the problem of separating the two effect curves in two ways: firstly by limiting the volume of normal tissue raised to a high dose, effecting a movement of the complication probability curve to the right (arrow 2); and secondly by producing a different dose in the tumour and normal tissues. The achievement of these two goals varies with anatomic site and application type. Gynaecological intracavitary insertions make extensive use of spacing and shielding to lower the dose to the normal tissues at risk, basically the bladder and rectum. That only parts of the organs receive high doses increases their tolerances. Many interstitial implants, on the other hand, must include the normal tissue in the treated volume, thus fixing the dose to the tumour and the normal tissue to the same value.

Few data exist for normal tissue tolerance. Most physicians develop their own feelings for what doses produce 'acceptable' probabilities for complications. In the absence of good values, and with the assumption that, for interstitial implants, the normal tissue coexists with the target in the implanted volume, some indication of tolerance follows uniformity of the dose in the treated volume, as discussed next.

3.2.1.3. Uniformity and comformality. In the 1930s, brachytherapy practitioners tried to establish some uniformity to implant techniques. Until that time, implant prescriptions simply referred to the strength of radium applied and the duration of the treatment. A physician may have varied the product of strength and time as a function of the implant size and/or shape, based on the clinician's experience. The actual amounts of radiation patients received varied greatly. The Manchester system first established the relationships between the implant geometry and the patient dose (albeit in units of exposure). Explicit in the relationships was the goal, noted above, that the dose through the implanted volume for volume implants or across the target plane for planar implants remains within $\pm 10\%$ of the nominal dose. While not stated in the paper describing the system, the justification for uniformity in the target lies with the joint requirements for delivering a necessary minimum dose to *all* tumour cells while maintaining the normal tissues in and around the tumour bed within tolerance. Since the normal tissue limited the maximum dose, delivering this dose to the entire volume gives the highest possible dose to all tumour cells. The Manchester system developed rules for the distribution of the radioactive materials to achieve uniform doses.

The approach taken in New York by Quimby differed markedly from that in Manchester. The Quimby system distributed radioactive material uniformly through an implant, often producing a much higher dose in the centre than at the periphery. The rationale for this dose distribution followed the argument that the centres of most tumours contain anoxic cells with reduced radiosensitivity. Therefore, the centre *requires* a higher dose for cure than does the periphery. The Quimby system prescribes to the minimum dose in the implanted volume (at the periphery) assuring that the entire target receives *at least* the prescribed dose. While few clinicians today admit to using the Quimby system, several subsequent systems—e.g., the Memorial, the Kwan (Kwan 1983) and the Zwicker (Zwicker 1985) systems—follow the uniform loading practice.

The Paris system, while using uniform loading through the implanted volume, adjusts catheter spacing based on the implant geometry, in part, to maintain some control on the maximum doses.

Thus, systems approached the question of uniformity very differently. In evaluating an implant for the uniformity of the dose, the physician first needs to decide on how uniform the dose across the target *should* be. Ling published some analyses suggesting that the ideal falls between the uniformity of the Manchester system and the high central dose of the Quimby system (Ling 1995). Such a conclusion could indicate an advantage for the Paris system.

3.2.2. Tools

3.2.2.1. Uniformity indicators. Evaluating the uniformity of an implant proves difficult at best because near the sources the doses approach incredibly large values. (Contrary to popular legions, the dose never approaches infinity unless the source is a singularity; the finite size and quantum nature of radiation keeps the doses in the realm of the ordinary.) Obviously, the determination of uniformity cannot simply rest on the difference between the highest calculated dose compared to the prescribed dose or the lowest dose in the target volume. Two very useful indicators of uniformity help in this evaluation.

Maximum significant dose

For any reasonable volume implant, the very high valued isodose surfaces conform to the sources, looking like the sources inflated. Isodose surfaces of just slightly lower values, for linear sources or seeds in ribbons, follow the needles or catheters, looking somewhat like cigars. As the values for the isodose surface decreases, the cigars grow in size, until at some isodose value two or more of the cigars coalesce. As seen in a plane passing through the union of the cigars, the isodose lines, which at higher values formed circles around single needle tracks, now encompass more than one needle. Very high doses in small volumes around single needles appear to be tolerated well by the body. Yet, a dose encompassing more than a single needle track can be significant with regards to bodily reactions. This isodose value where the cigars coalesce, the highest isodose value to encompass more than a single needle, defines the *maximum significant dose*, or MSD.

By strict application of the definition, the MSD would appear to enclose two needles only in one plane, that with the highest doses. Other planes would continue to show the isodose lines corresponding to that value as circles around the needle tracks. In our practice, we cheat on the application of the definition a bit, selecting the isodose surface that either encompasses three needle tracks in a plane, or at least two sets of two needles. In general, for implants of 50 cm^3 or greater, avoiding serious complications requires maintaining the MSD less than 10% above the prescribed isodose value, assuming that the latter surrounds the implanted volume as with the Paris system (Fleming *et al* 1984).

Minimum contiguous volume

A concept related to the maximum significant dose is the minimum contiguous volume. Continuing to reduce the value for the isodose surface from the MSD, the surfaces encompass increasing amounts of the sources. At some point, the isodose surface encompasses all of the radioactive material. The volume enclosed by this surface is called the minimum contiguous volume (MCV) and the isodose value the maximum contiguous dose (MCD) (Neblett *et al* 1985).

3.2.2.2. Volume–dose histogram. For the decades before computer calculations, analysis of the 'quality' of an implant used little more than visual assessment of how closely the needles approximated the plan. Calculation of the nominal dose for the executed implant geometry provided a numerical comparison to the ideal. Computerized calculations that supplied isodose lines changed markedly the standards for evaluation of implants, allowing clinicians to see whether the isodose line corresponding to the dose rate prescribed covered the intended volume, and whether the higher valued lines took in too much tissue. Even if not quantitative, the qualitative evaluation improved markedly. Yet, there lingered a desire for a more objective, numerical method for analysis. The volume–dose histogram (VDH) addresses this desire[1].

Basic volume–dose histogram. The simplest form of the volume–dose histogram plots as a bar graph the volume enclosed between specified isodose surfaces. An example helps demonstrate the concept. Figure 3.5 shows an implant used to deliver a boost dose to a tumour in a prostate following external-beam treatment, including the implant and isodose lines superimposed on several CT images for this patient. The isodose lines in this example display dose rates. The assessment of the volume between adjacent isodose lines may integrate volumes using interpolated isodose surfaces or simply add pixels in each plane, and assign them a volume

[1] The volume–dose histogram is also known, possibly more widely, as the dose–volume histogram (DVH). Properly, since the volume occupies the ordinate, the former term should apply. Exceptions occur for historical quantities in existence before nomenclature conventions, or, though named contrary to convention, in extremely widespread use. Although depth-dose curves are held up as a contraconventional example, the name applies not to curves of dose against depth, but of the quantity depth-dose (i.e., dose at depth divided by dose at peak) against depth, with only the ordinate stated.

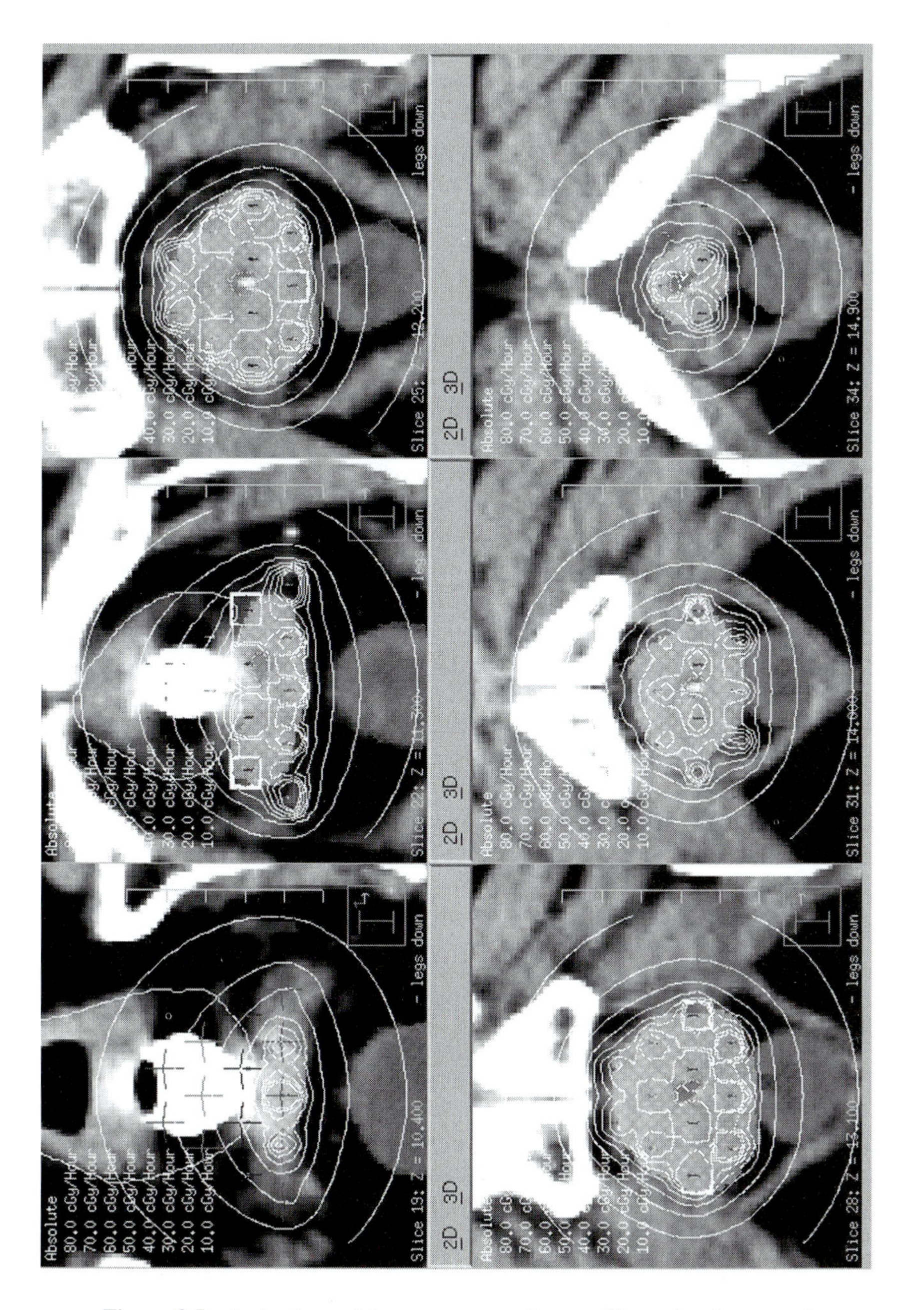

Figure 3.5. *An implant of the prostate, used as an illustration for several evaluation quantities.*

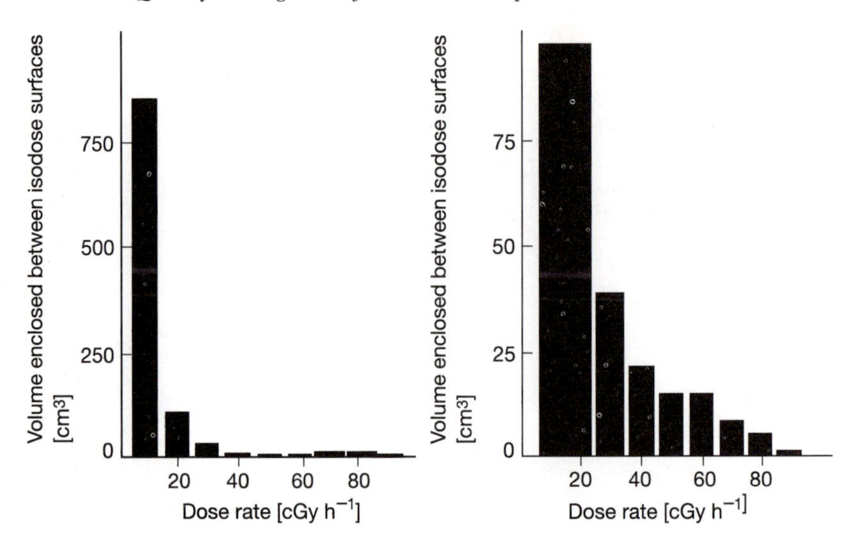

Figure 3.6. *The basic volume–dose histogram for the implant in figure 3.5. The graph on the left simply shows the same information as that on the right, except with a smaller scale to highlight the slight plateau around the treatment dose rates.*

based on the pixel size and the slice thickness. Figure 3.6 shows the basic or simple VDH for this implant as the black boxes. The value for the lowest interval, 0 to 0.1 Gy h^{-1}, for many computer systems includes the universe outside the 0.1 Gy h^{-1} surface, and may limit to infinity. The value for this interval rarely becomes important and may be deleted without much loss of information. More sophisticated programs stop at the patient surface. These simple volume–dose histograms for volume implants generally exhibit similar characteristics:

- very large volumes for very low dose rates,
- very small volumes for very high dose rates and
- a slight peak or plateau somewhere in the middle in the neighbourhood just higher than the isodose surface that encircles much of the implant.

These characteristics each make sense. The low dose rate isodose surfaces fall far from the sources, thus enclosing large volumes. Coincidentally, as a product of the inverse square relationship between dose rate and distance, the gradient between surfaces becomes less with distance, so the space between these large surfaces becomes increasingly larger as the dose rates decrease. In contrast, the high dose rate surfaces wind tightly around the sources, with high gradients between, so the volumes enclosed become small. The plateau comes by construction of the implant, through the attempt to produce a uniform dose rate region near the prescribed dose rate. A simple VDH for a treatment for cancer of the cervix with a tandem and ovoid exhibits no peak or plateau since the sources all reside in the

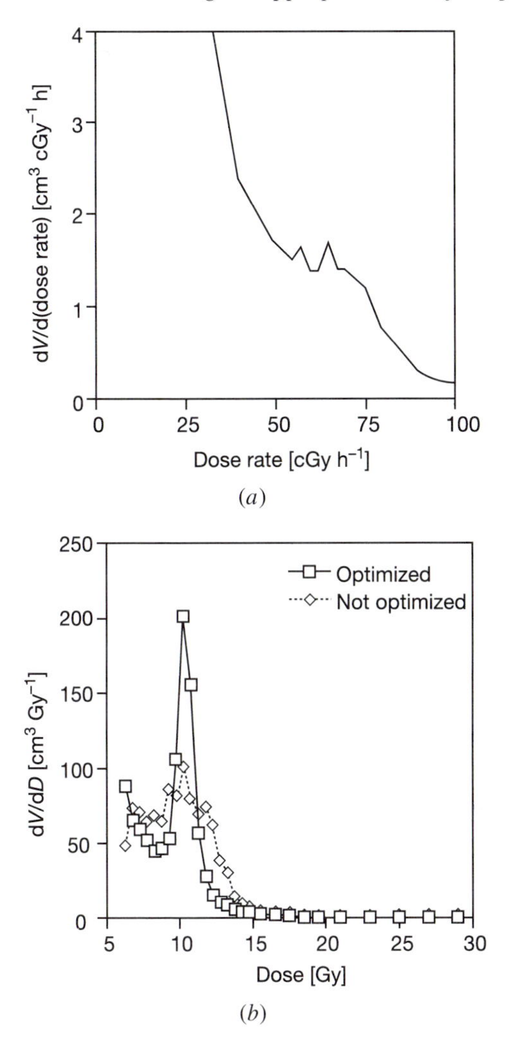

Figure 3.7. *(a) The differential volume–dose histogram for the prostate implant in figure 3.5. (b) Differential VDHs for two cylindrical volume implants, identical except that one was optimized. The bunching of the dose rates around the target dose becomes more apparent for the optimized case.*

middle of the target volume with the dose rates continually falling with increasing distance.

Differential volume–dose histogram. Even though the VDH for the volume implant in figure 3.5 contains a plateau, this feature stands out poorly from the general shape of the histogram. Smaller intervals for the dose rates help to highlight some of the details of the distribution. Extending this process to extremes, the continually decreasing size of the Δ (dose rate) becomes the differential d(dose rate). For the simple VDH, the sum over all of the columns gives the total volume of the patient. The summation becomes analogous to integration with the ordinate seen as the volume per dose interval, or Δ(volume)/Δ(dose rate), except that the negative sign is usually ignored. When reducing the dose rate bins to infinitesimals, the integral under the resulting figure should retain the equivalence with the volume of the patient. Thus, the quantity plotted on the ordinate becomes not d(volume), but d(volume)/d(dose rate), so that

$$\text{total volume} = \int_{all\ dose\ rates} \left(\frac{\mathrm{d(volume)}}{\mathrm{d(dose\ rate)}} \right) \mathrm{d(dose\ rate)}.$$

A plot of d(volume)/d(dose rate) as a function of d(dose rate) displays a *differential* volume–dose histogram. Figure 3.7 shows the result of this process for the example implant. The figure also plots two curves, one with the distribution from a large, cylindrical volume implant optimized to deliver a relatively uniform dose through the target volume, and the other with the same geometry but using uniform distribution of equal-strength sources. Both distributions deliver the same dose rate to the periphery of the implanted volume. The bunching of volume into the dose rates just above the target dose rate becomes obvious in the optimized case, and the size of this 'peaking' compared to the non-optimized curve often serves as an indicator of quality for an implant, as discussed below in section 3.2.2.3.

Cumulative or integral volume–dose histogram. A different way of looking at the same information plots the total volume enclosed by a dose rate isodose surface. Because the isodose values decrease with increasing distance from a source, for any dose rate this becomes equivalent to plotting the volume that receives at least that dose rate. Such a graph is called a *cumulative* or *integral* volume–dose histogram. For our example implant, figure 3.8 shows the cumulative VDH, with several features common to all such curves:

- the curve constantly decreases as the dose rate increases;
- optimized and uniformly loaded curves exhibit much less of a difference;
- an optimized curve displays no peak (as a result of the first-listed characteristic), and only a slight, drooping plateau.

For a cumulative VDH calculated without a patient boundary (i.e., in an infinite medium), the curve limits to infinity towards very small dose rates using conventional calculational algorithms, just as discussed above for a simple VDH. Considering the patient surface as the edge of the calculational volume maintains these lower dose rate values to physical dimensions.

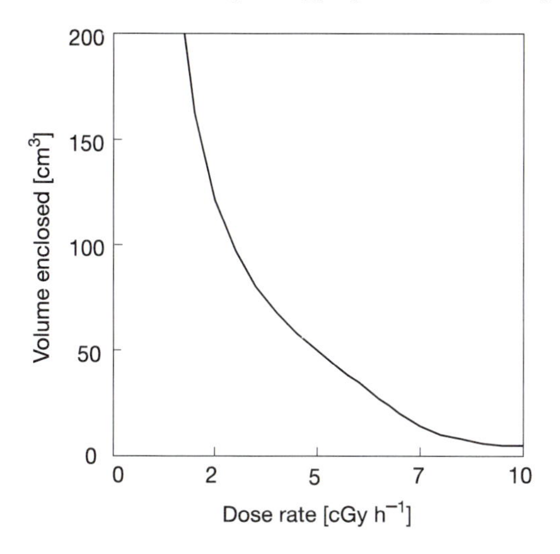

Figure 3.8. *The cumulative volume–dose histogram for the prostate implant in figure 3.5. Notice the lack of features compared with the differential VDH.*

Natural volume–dose histogram. Anderson (1986) developed a special modification of the differential VDH that assists in evaluation of the extent of uniformity in an implant, called the *natural volume–dose histogram* (NVDH). Outside the implanted volume, the distribution of dose rates begins to look much like it would for a point source at the centre of the volume that contained all of the activity. The lower the dose rate values, the more the curve assumes the shape of a point source's distribution. This overall superimposed pattern partially masks the structured detail produced by the distributive geometry of the implant. The NVDH seeks to remove the simple inverse-square law part of the dose distribution from the histogram's display. The first step to this removal recognizes that the volume, V, of a sphere around a point source relates to radius, r, as

$$V = \tfrac{4}{3}\pi r^3$$

and that, for a point source,

$$D = \frac{S_K k}{r^2} \Rightarrow r^3 = \left(\frac{S_K k}{D}\right)^{3/2}$$

where

$D =$ the dose at distance r,

$S_K =$ the air kerma strength of the source, and

$k =$ constants that relate the source strength to dose.

Substituting the equation for dose into that for volume gives

$$V = \tfrac{4}{3}\pi(S_K k)^{3/2} D^{-3/2}.$$

Differentiating this equation yields

$$\frac{\mathrm{d}V}{\mathrm{d}D} = 2\pi(S_K k)^{3/2} D^{-5/2}$$

for a point source. Anderson proposed differentiating not with respect to D but with respect to the function $u(D) = D^{-3/2}$. This derivative becomes

$$\frac{\mathrm{d}V}{\mathrm{d}u} = 2\pi(S_K k)^{3/2}$$

a function independent of dose, but only for a point source. Plotting a histogram of $\mathrm{d}V/\mathrm{d}u$ as a function of dose produces a horizontal line for a point source, while deviations from a point-source type distribution appear as non-horizontal features. Figure 3.9 shows the NVDH for the large, cylindrical example implant, both with uniform loading and optimized. Both cases significantly differ from a straight line, but the optimized treatment displays a much more prominent peak. The lower dose side corresponds to distances far from the implant, at which locations the source distribution differs little from a point source. On the high dose side, the distributions again look like those very near point sources, but in the case of the distributed source, the intensity is much lower than where all the activity is in a single point, so the curve falls below that on the left.

Limited volume–dose histograms. Both of the VDH types discussed above refer to volumes found anywhere through the entire patient for any value of dose rate (or dose if integrated over a treatment time). Such VDHs are termed *holistic*. Alternatively, often the statistics for specific organs form the quantities of interest. Figure 3.10 shows the cumulative VDH for the prostate example implant considered before, but in this case, for each dose rate on the abscissa, the ordinate gives the fraction of a designated region of interest raised to at least that dose rate. For this case, 100% of the target falls within the 0.38 Gy h^{-1} isodose surface. As the dose increases, the fraction of the target included decreases because the isodose surfaces begin to withdraw into the implant and, thus, into the target. The isodose curves tell *where* the volumes of lower dose occur; the VDH does not. Eventually, at high dose rates, the volumes become very small. Depending on the behaviour of the calculational algorithm at very short distances, the volumes will fall to zero at some cutoff value of dose rate, or will asymptotically approach the abscissa. The true situation falls between. The interest in such a display follows the assumption that for cure, all of the cells in the target must receive some minimum dose. Some biological models become more complicated, hypothesizing that some fraction, such as 85%, must receive some higher dose. A VDH as shown displays the relevant information.

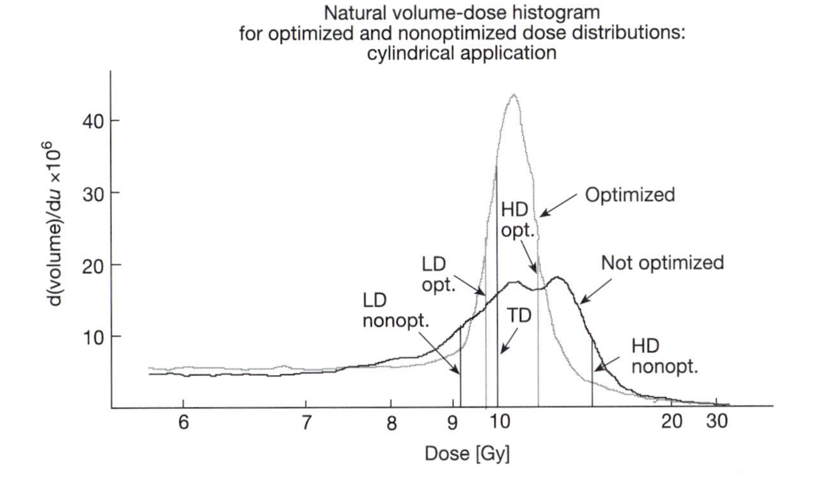

Figure 3.9. *Natural volume–dose histograms for the large, cylindrical volume implants. In the figure, LD represents the dose for which the* dV/du *falls half way between the peak value and the background level to the left. HD represents the corresponding quantity on the high dose rate. TD equals the target dose.*

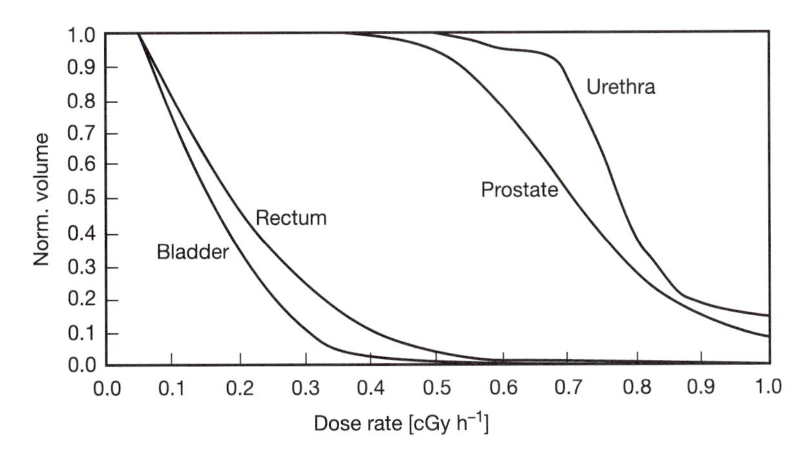

Figure 3.10. *A limited, cumulative VDH for the prostate implant. For each dose rate, the ordinate gives the fraction of the region of interest raised to that level. In this implant, the high dose rates to the urethra could be a cause for concern, depending on the total dose prescribed and previous radiation history.*

The graph also displays analogous curves for the bladder, rectum and urethra. In these cases, the concern turns to tolerance of the dose received. Most biological models consider the dose an organ tolerates to be a function of the volume of

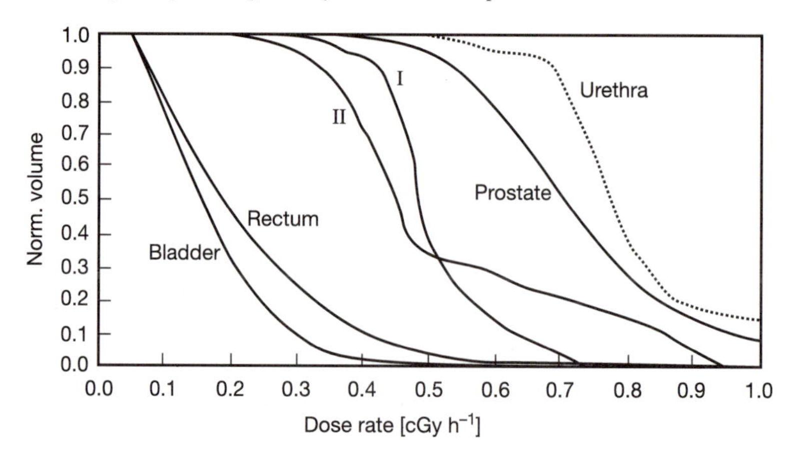

Figure 3.11. *Two hypothetical VDH curves for an organ: I, which exposes larger volumes to low doses, and II, which exposes smaller volumes but some to a higher dose.*

the organ raised to that dose. Analysis of expected tolerance based on these curves becomes problematic for two reasons. Firstly, the computer only considers the organ volume included in the CT study. For an organ such as the rectum, most of the volume exists outside the CT range. Thus, the ordinate value for 'fraction of the contoured volume' deviates markedly from the fraction of the organ. Secondly, the developers of biological models, as of this writing, understand poorly the variables involved. As an example, consider figure 3.11, which shows two hypothetical VDHs for an organ. The regimen for curve I exposes more cells to low doses than that for curve II, but the curve II treatment takes some small volumes to much higher doses. Which curve represents the treatment with the lesser probability for complication depends on the biological model used. Notwithstanding such complex situations, these VDHs often clearly indicate which of the possible treatment plans delivers the best distribution to the target with the least likelihood of complication.

Because the histograms in figure 3.10 normalize the volumes within dose rate surfaces to the volume of the contoured region of interest, they are called *normalized, limited* VDHs, limited to the region of interest. Characteristically, normalized, limited cumulative VDHs have flat tops, with a fractional value of 1.0 from very low dose rates to some value just failing to cover the contour. From that dose rate to higher values, the curve continually decreases.

Limited VDHs also can be differential, as figure 3.12 displays. This depiction of the VDH gives information about the distribution of doses in a region of interest not seen in the cumulative limited VDH.

In summary, volume–dose histograms can be either holistic, looking at the implant as a whole without regard to any particular regions of interest, or lim-

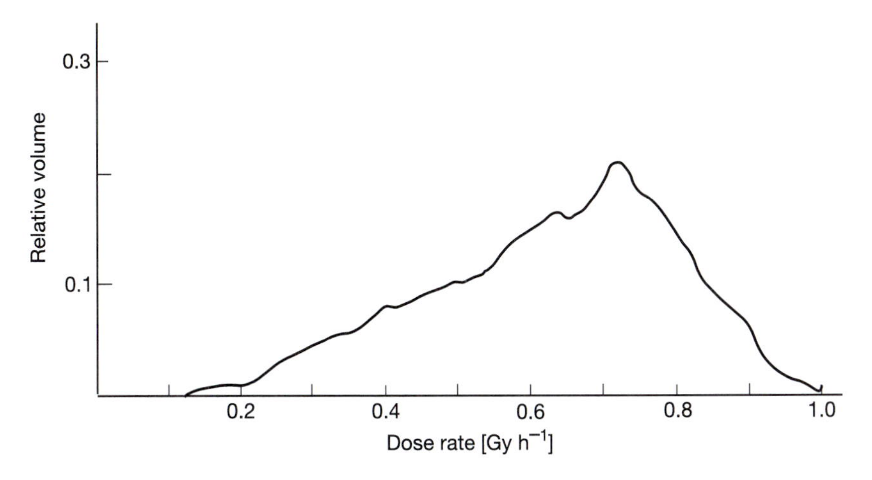

Figure 3.12. *The differential, limited, normalized VDH for the prostate in the example implant in figure 3.5.*

ited, looking only at a region of interest. Both holistic and limited VDHs can present the information as cumulative, showing the total volume enclosed by a given isodose surface, or differential, plotting the change in volume at each isodose surface. The limited VDHs may normalize the volumes to a fraction of the total volume of a region of interest, or use the absolute volumes. Each of the types of VDH present different information. As noted, the limited, cumulative VDH helps assess the adequacy of target coverage and tolerance of normal structures. The differential, holistic VDH conveys information about the quality of the implant without regard to the patient. While the latter may seem to miss the point, in practice this often carries useful information for prescribing the treatment, following the assumption that the implant borders approximate the target volume. However, that discussion requires some more tools, as discussed next.

3.2.2.3. Indices. In this discussion of evaluating brachytherapy treatment plans, an *index* refers to a numerical indicator of quality for a specified aspect of the treatment plan. All of the indices discussed here have some physical meaning, although through various normalizations, sometimes the meaning becomes obscured, and they seem arbitrary numbers. Most often in practice, their exact meanings become relatively unimportant because their values either should fall within an accepted range or serve as a tool for ranking competitive plans.

Many of the indices make use of some of the following commonly defined quantities. These definitions, as with all that follow, use dose and dose rate interchangeably unless noted. In general they differ only by a constant.

MCD: the mean central dose
The arithmetic mean of the local minimum doses between all adjacent sources in the implant. The MCD corresponds approximately to the basal dose of the Paris system.

PD: peripheral dose
The minimum dose at the periphery of the clinical target volume.

TD: target dose, or the prescribed dose
The dose nominally intended to be delivered to the target volume. Usually, this is thought of as a minimum dose to the target volume, but that is not always the case. Several systems specify the target dose as other than the minimum dose to the target volume. However, most modern practitioners have turned to pairing the target dose with target volume, at least in the ideal.

HD: high dose
A dose significantly higher than the target dose, of concern because of complications. The definition of 'significantly' varies between authors. The Paris system defines the high dose volume as containing doses exceeding the target dose by a factor of 2, while Saw and Suntharalingam (1991) and the ICRU (1997) use 1.5. Zwicker and Schmidt-Ullrich (1995) make the factor a variable, p, adjusted as part of the optimization process. Anderson (1986) defined the quantity based on the natural VDH, as the dose rate half way between the peak dose rate and the baseline formed by a simple inverse square distribution, toward the higher dose rate side. Using a natural VDH simplifies finding the baseline because it forms a horizontal line. HD is indicated on figure 3.9. Such a definition only applies to interstitial implants, since intracavitary insertions seldom exhibit peaks in their VDH.

LD: low dose
A dose significantly lower than the target dose, of concern because of potential treatment failure. While not quite as controversial as HD, various values have been suggested. A commonly used value is 90% of the TD. The ICRU specifies 90% of the PD for the LD. As with the HD, Anderson (1986) uses the value half way between the peak and the baseline on the lower dose rate side. Figure 3.9 also shows the LD.

Each of the specified doses above have a corresponding volume enclosed receiving at least that dose, and indicated by the appending of a 'V' to the quantity. For example, *HDV* denotes the volume raised to at least the high dose, defined as the *high dose volume*.

The ICRU (1993) defined several volumes of interest in radiotherapy:

GTV: gross tumour volume
The three-dimensional volume in the patient bounded by the observable tumour.

CTV: clinical target volume
The GTV plus a margin accounting for microscopic invasion or other non-visible

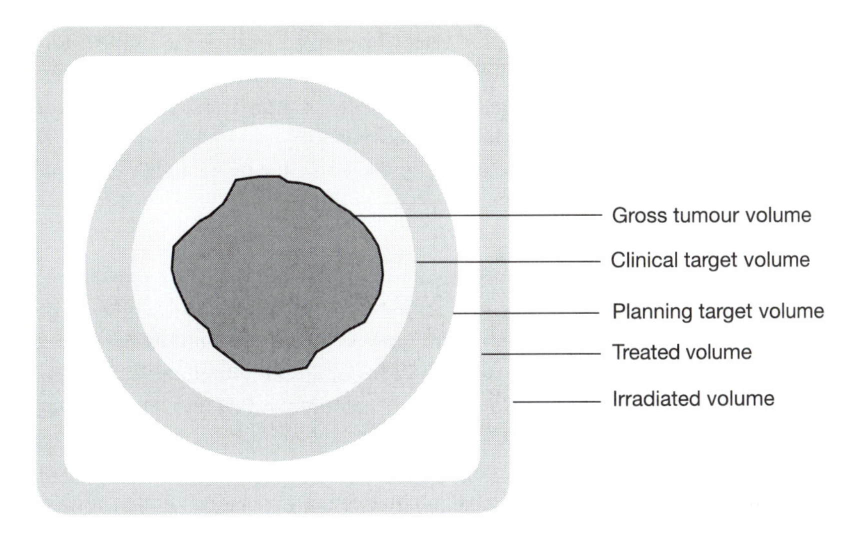

Figure 3.13. *Illustrations of the conceptual 'target volumes'. (Figure from Thomadsen and Hendee 1999. Figure used with the permission of VSC-Wiley Publishing.)*

extension of the tumour cells into normal tissue. This margin depends on the nature of the tumour and the surrounding normal tissue.

PTV: planning target volume
The CTV plus a margin accounting for treatment uncertainties. In brachytherapy, such uncertainties include the inability to ensure insertion of the appliance into the planned location or geometry.

TV: treatment volume
The volume in the actual treatment raised to the prescribed dose or above.

Figure 3.13 illustrates these volumes. Ideally, the target dose volume would equal the PTV. However, in real implants this identity seldom, if ever holds. The treatment volume, as opposed to the previous three quantities, applies postexecution, and essentially equals the target dose volume, although the target dose volume sometimes refers to a planned application, and other times to that executed.

The indices of the first set each indicate a single aspect of an implant related to target volumes.

CI: coverage index
The fraction of the target volume receiving a dose equal to or greater than the target dose. The CI corresponds to the value on the relative, cumulative VDH for the target dose. Obviously, the ideal value equals 1.0. The definition fails to distinguish the type of target volume under consideration, be it gross tumour volume, clinical target volume, planning target volume or treatment volume. When

originated, these distinctions had not been formalized. Since the clinical target volume supposedly contains target cells, the dose to this volume should be of interest. That the planning volume has a margin to assure coverage of the clinical target volume indicates that the latter is the quantity of interest. Thus, in this discussion, the CTV will serve in calculating the coverage index and others, unless noted.

EVI: external volume index
The volume of tissue outside of the target volume receiving doses equal to or greater than the target dose, as a fraction of the target volume. In the literature, it often occurs as EI. Again, because the definition predates the distinctions between the various tumour and target volumes, the EVI becomes a bit ambiguous in modern terms. The most reasonable interpretation of the intention of this quantity would use the clinical target volume in the definition for *the* target volume. This would then indicate the 'price' paid in terms of normal tissue, for the coverage of the clinical target. In part, such an analysis includes the measure of expansion from the clinical target volume to the planning target volume, as well as the further increase necessary in the treatment volume.

RDHI: relative dose homogeneity index
The fraction of the target volume receiving a dose between the target dose and the high dose. Some references simply use the term homogeneity index, HI. In this case, the intention of the definition does not clearly indicate which volume to pick for the target volume. Were this quantity to correlate with the probability of a cure without complication, the clinical target volume would make the most sense. If, on the other hand, the intention were to rate the merits of the plan, or its execution, then 'target' should read planning target volume. For the most part, the use of the clinical target volume probably would prove most useful, but either may find applications, and may be distinguished by subscripts, as $RDHI_c$ and $RDHI_p$, respectively.

DNR: nonuniformity ratio
The ratio of the high dose volume to that volume taken to at least the target dose, i.e.,

$$\mathrm{DNR} = \frac{\mathrm{HDV}}{\mathrm{TDV}}.$$

The intent is that this quantity relates to the probability for complication. This intent implies the use of the actual treatment volume in the denominator. The target dose volume, in contrast, forms an important decision variable during planning. The DNR serves as one measure of the effects of a choice in target dose.

Saw and Suntharalingam (1991) compared the values for these last four indices for a given implant as a function of the selected target dose rate. Figure 3.14 is redrawn from their study. As the figure shows, coverage, as indicated by CI, is approximately full below 40 on the abscissa, while homogeneity (HI) peaks at around 43 to 44. The DNR indicates a minimum at approximately 50, the same

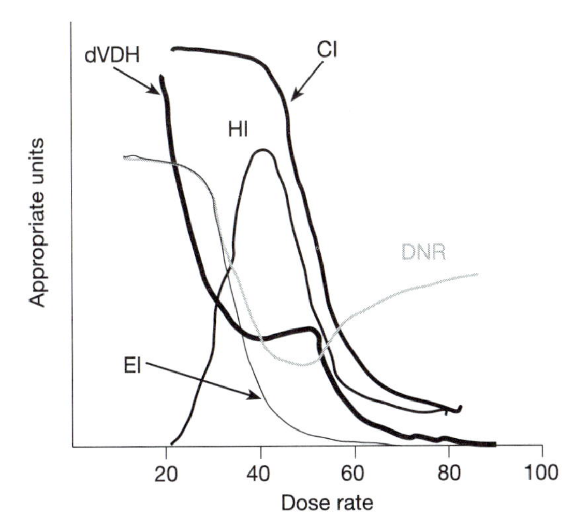

Figure 3.14. *A comparison of the values for several index quantities from Saw and Suntharalingam (1991). (Figure from Thomadsen et al 1997. Figure used with the permission of Futura.)*

value as the external volume index shows as a minimum to avoid appreciable exposure out of the target. The fact that the indices produce different optimum values should not be a surprise, since each looks only at a given facet of the implant. No one number tells the whole story.

Other indices relate to single aspects of the implant, but without regard to any regions of interest.

Spread

The difference between the highest and the lowest local minima in the implanted volume divided by the mean central dose. This quantity gives an indication of the appropriateness of needle track spacing.

PMR: peripheral–mean ratio

The ratio of the peripheral dose to the mean central dose, or

$$\mathrm{PMR} = \frac{\mathrm{PD}}{\mathrm{MCD}}.$$

Excessively low values for the PMR indicate an attempt to reach too far outside the physical implantation.

UI: uniformity index

A measure of dose uniformity over the treatment volume, defined as

$$\mathrm{UI} = \frac{\mathrm{TDV} - \mathrm{HDV}}{\mathrm{TD}^{-3/2} - \mathrm{HD}^{-3/2}} \Bigg/ \frac{\mathrm{TDV}}{\mathrm{TD}^{-3/2}}.$$

The uniformity index depends on the target dose selected not only through the explicit entry of target dose quantities, but also through the dependence of HD on the TD. Zwicker notes that UI also depends on the value for p used in defining HD. The dependence of HD on TD disappears when defining HD based on the width at half the maximum for the natural VDH, as discussed above. While dependent on the target dose, this index contains no reference to any target volume, nor the uniformity across that volume. Use of the index assumes that the target dose matches well the target volume of interest. The use of the $-3/2$ power in the exponents in the denominators makes the equation equivalent to

$$\mathrm{UI} = \frac{\mathrm{TDV} - \mathrm{HDV}}{u(\mathrm{TD}) - u(\mathrm{HD})} \Bigg/ \frac{\mathrm{TDV}}{u(\mathrm{TD})}.$$

As with the natural VDH, the intention seems to be to measure uniformity as improvement over the distribution produced by a single point source. For a point source, $\mathrm{UI} = 1$ identically.

The ratio of the numerators gives the useful volume of the implant as a fraction of the total target dose volume. This ratio increases as the useful volume of the implant increases. The fraction as a whole displays how this fractional useful volume for the implant compares with that of a point source at the centre of the implant.

Unfortunately, the maximum value for the uniformity index does not always indicate the most appropriate target dose rate for an implant. For the prostate implant in figure 3.5, the target dose rate that covers the prostate is 0.5 cGy h^{-1}. However, the UI increases from a value of 1.48 for a TD of this dose rate to a maximum of 2.57 at 0.9 cGy h^{-1}. As the target dose rate increases, the HDV decreases faster than the TDV until the isodose surfaces start wrapping around the sources. While this index shows promise, it will require further refinement before wide application.

QI: quality index

A measure of the uniformity of the implant without regard to the target dose, defined as

$$\mathrm{QI} = \frac{\mathrm{LDV} - \mathrm{HDV}}{\mathrm{LD}^{-3/2} - \mathrm{HD}^{-3/2}} \Bigg/ \frac{\mathrm{LDV}}{\mathrm{LD}^{-3/2}}$$

or equivalently

$$\mathrm{QI} = \frac{\mathrm{LDV} - \mathrm{HDV}}{u(\mathrm{LD}) - u(\mathrm{HD})} \Bigg/ \frac{\mathrm{LDV}}{u(\mathrm{LD})}.$$

While not dependent explicitly on the target dose, the quality index varies as a function of TD through HD and LD unless they come from the width at half the maximum of the natural volume–dose histogram. The developers of this index intended the latter definitions for HD and LD to free this index from the selection of TD so it provides an independent measure of the implant ‘quality’. Again, for a point source, the QI identically equals 1.

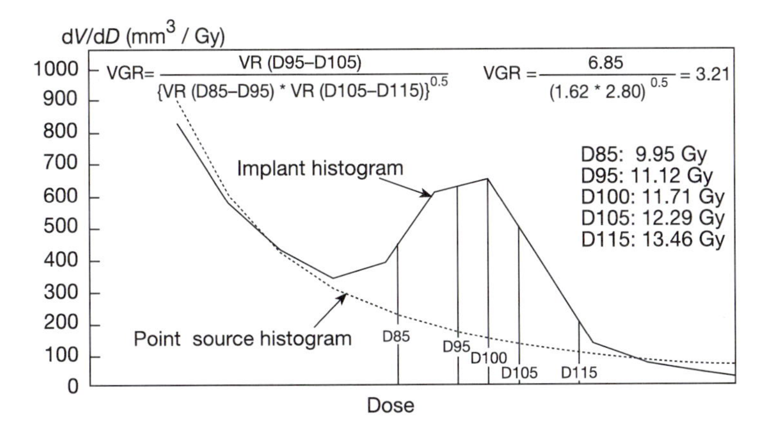

Figure 3.15. *An example for the determination of the volume gradient ratio (VGR) from a differential histogram. As defined in this figure (with the peak value as 100%), the VGR would give an indication of the homogeneity of the dose in the implanted volume, but not necessarily the homogeneity given the target dose.*

VGR: volume gradient ratio

A measure of the uniformity of the target dose for an implant compared to that from a point source. The definitional equation begins by defining *the volume ratio*, VR. To find the volume ratio, plot the differential VDH for the implant and for a point source of the same total strength together, as in figure 3.15. Assign the target dose rate as shown the arbitrary value of 100%. The $\mathrm{VR}(D_{95} - D_{105})$ is the volume between the 95% dose rate and the 105% dose rate for the implant, divided by the volume between those same dose rates for the point source. The volume gradient ratio becomes

$$\mathrm{VGR} = \frac{\mathrm{VR}(D_{95} - D_{105})}{\sqrt{\mathrm{VR}(D_{85} - D_{95})\mathrm{VR}(D_{105} - D_{115})}}.$$

The VGR specifies the relative improvement the implant exhibits around the target dose rate over a simple point-source distribution, normalized to the geometric average of the improvement for the 10% dose rate intervals above and below the target dose rate. In concept, the target dose rate should cover a large volume, but ideally, the volumes should diminish quickly outside of the target volume, so the VR in the denominator should tend toward unity. Large values for the VGR indicate uniform doses through the target volume with small high dose volumes and rapid decreases in dose rate outside the implanted volume. The selection of the symmetric 95% and 105% dose rate for the numerator (and the corresponding values in the denominator) may not always be appropriate. For implants using the minimum peripheral dose for the target dose, the following modified version may relate more to the application:

$$\mathrm{VGR}' = \frac{\mathrm{VR}(D_{100} - D_{110})}{\sqrt{\mathrm{VR}(D_{90} - D_{100})\mathrm{VR}(D_{110} - D_{120})}}.$$

3.3. SUMMARY

Evaluating a treatment plan before execution consists of assessing two basic aspects of the plan:

(1) the correctness of the plan,
(2) the appropriateness of the plan.

Checking the correctness of the plan principally looks for errors that may have crept in through the many steps of the planning process. The list below suggests items to assess.

(i) Direct checks on the input data, checking:
  (a) source strengths,
  (b) appliance used,
  (c) dose prescription and specification and
  (d) dose rate and treatment duration for temporary applications.
(ii) Whether the plan make sense.
(iii) Indirect checks on the input data, such as:
  (a) calculation of the dose rate to selected points or
  (b) calculation of the dose distribution using a second planning system.

The evaluation of the appropriateness of the plan assesses:

(i) coverage of the target;
(ii) tolerance of sensitive, normal structures nearby and
(iii) uniformity and conformality of the dose distribution.

Some tools to assist in this evaluation include:

(i) uniformity indicators, such as maximum significant dose and minimum contiguous volume;
(ii) volume–dose histograms, relative or absolute, and cumulative or differential (including the 'natural'); and
(iii) indices, such as
  (a) coverage index,
  (b) external volume index,
  (c) relative dose homogeneity index,
  (d) dose nonuniformity ratio,
  (e) peripheral–mean ratio,
  (f) uniformity index,

(g) quality index and
(h) volume gradient ratio.

Any of the tools used in assessing a treatment plan that consolidate the three-dimensional information into a more compact form, whether a two-dimensional volume–dose histogram, or a single-valued index, lose some important information along the way. The isodose surfaces, and particularly the two-dimensional superposition of the surfaces as isodose lines on image planes, provide the most information, and the final assessment of adequacy of a treatment plan.

CHAPTER 4

QUALITY MANAGEMENT FOR MANUALLY LOADED, LOW DOSE RATE APPLICATIONS

Much of the review necessary for low dose rate (LDR) treatment centres on the correctness of the treatment plan, a topic discussed in a separate chapter. There remain, however, many possible paths for error that need closing. The potential pitfalls depend strongly on the nature of the application. The sections below cover several of the more common types of treatment, although hybrid varieties or new modalities may be encountered. The general principles should pertain.

4.1. APPLICATIONS INVOLVING SOURCES FROM A STANDING INVENTORY

4.1.1. Source loading

Gynaecological intracavitary insertions using ^{137}Cs form the most common examples of treatments using sources taken from a standing inventory. As discussed in chapter 3, one of the last parts of the review includes a check that the correct sources occupy the proper positions. Seeing that such applications use relatively few sources, checking each source individually takes little time. A single source with the wrong strength easily produces errors in dose exceeding 20%. The manufacturers of intracavitary sources have used different approaches to marking their strength. Table 4.1 summarizes the methods used, and each has benefits and disadvantages. Most modern afterloading appliances use one or both of the means employed by the Fletcher–Suit applicator for treatment of cancer of the cervix: a uterine tandem and vaginal ovoids. For this applicator, the sources in the tandem are first slid down a clear plastic tube (sealed at one end), in the source-storage facility. Later, in the patient's room, the person loading inserts the plastic tube containing the source into the hollow tandem previously placed into the patient's uterine canal. For this type of loading, colour coding on the side of the source

Table 4.1. *Some examples of identification techniques for intracavitary sources.*

Manufacturer	Method of coding	Code for sources of 25 mg Ra eq	20 mg Ra eq	15 mg Ra eq	10 mg Ra eq
US Radium	Tint on the capsule	All silver	2/3 silver, 1/3 gold	1/3 silver, 2/3 gold	All gold
Amersham	Black rings on the side	5 rings	4 rings	3 rings	2 rings
Nuclear Associates	Coloured bands on the side[a]	White	Blue	Black	Red
3M	Colour on the eyelet end	White	Blue	Black	Red

[a] The sources carry three coloured bands, indicating the source strength, year of manufacture and source model. The person checking the source loading must understand the order of the bands to avoid confusion of the other bands with that for source strength.

tube makes checking the strength of the source in each position just before loading an easy matter. However, the sources themselves and any stylet used to hold the sources in place obscure marks on the ends of the tubes.

Loading for the Fletcher–Suit ovoids proceeds very differently. Because the final orientation of the sources lies almost perpendicular to the handles (and access channels) of the ovoids, the sources sit in metal buckets on double hinges. Once in the bucket, a colour code on the top of the source remains visible, while that on the side becomes hidden. Thus a single marking system fails to provide indication of source strength in one or the other part of the applicator. A common attempt at solving this dichotomy finds the physicist adding enamel colour codes to either the top or side of each source. By choice, manufacturers produce sources with nonreactive jackets, and user-added marks require frequent refreshing, particularly on the side, where the source rubs against the appliance. User-added enamel marks on the side sometimes also cause problems fitting the sources into holders without much clearance. The manufacturer colour codes on the side use enamel-impregnated porcelain laid into machined channels to yield a uniform diameter along the source length. However, with heavy use and time, even these manufacturer markings fade or crack and fall out of their tracks.

Any department that has added sources to its inventory most likely has sources with the same nominal strength but different true strengths. In such cases, the additional colour-coded date of manufacture, as used with the Nuclear Associate brand sources, allows visual differentiation between the batches. Often, such additions come when the original sources decay to the point that treatment times become excessively long (very low treatment dose rates compromise the biological effectiveness, or third-party payers object to the extra day in the hospital). Replacing the entire set of sources proves an expensive undertaking, so many facilities simply replace the strongest sources (usually 25 mg Ra eq) with new, and redesignate the older 25 mg Ra eq sources as nominal 20 mg Ra eq. The old 20 mg Ra eq sources replace the 15 mg Ra eq, and the 15 mg Ra eq move to the 10 mg Ra eq position. The old 10 mg Ra eq often assume the role of 5 mg Ra eq, a strength seldom ordered originally. Two problems follow from this practice:

(1) The ratio of strengths for the sources changes from the original set. Assume that the strengths of the original sources actually came in the ratio of 25:20:15:10 (or 5:4:3:2), even though at delivery the actual strengths should have exceeded their nominal value. Using sources with the nominal strength often gives a dose rate of 0.55 Gy h^{-1} at a target point. When the actual dose rate falls to 0.44 Gy h^{-1} with the same sources, the strengths will be 20:16:12:8. Further assuming that the purchaser would wish to extend the useful life of the new sources by buying a source with higher than nominal strength, often 10% higher such as a 27.5 mg Ra eq, the set follows the ratios of 5.5:4:3.2:2.4. Simply using the same nominal loadings as with the original set can produce very different dose distributions, particularly if the source loading uses several 10 and 20 mg Ra eq sources together. Calculating treatments using the true strengths eliminates errors in dose, although the distributions will differ from those previously used.

(2) The marking of the sources may cause confusion. Aside from the aspects of this problem discussed above (different strength sources with the same colour code), great care must be taken to assure that all users know that the designation corresponding to a given colour has changed.

After loading the sources into the patient, a visual inspection of the area around the applicator for sources that may have slipped out of the appliance during the loading procedure takes little time but can prevent tragedies. Particularly with the Fletcher–Suit ovoids, several sources have fallen out of the holder as the person loading wiggled the holders trying to get them into the ovoid handles. Unnoticed, these sources can cause skin burns if the source lodges against the patient, or can expose many unsuspecting persons if folded into an absorbent pad under the patient and tossed into the rubbish.

4.1.2. Calculation of the removal time

Calculating the time to remove the sources (removal time, or sometimes the 'out time') usually requires three pieces of information: the target dose, the target dose rate and the time of insertion ('in time'). That the sources come from a permanent inventory implies half-lives very much greater than the duration of application, so decay *in situ* can be ignored. The target dose rate, of course, comes from the treatment plan, and considerations involving that fall in chapter 3. The treatment duration simply becomes

$$\text{duration} = \frac{\text{dose}}{\text{dose rate}}. \tag{4.1}$$

Because loading takes approximately 30 s, precision in ascertaining the insertion time proves no problem. The greatest problem becomes actually noting the time. However, since a normal treatment lasts about 60 h, it takes a 3 h error to cause a 5% error in dose.

The most likely causes of significant error in calculating the removal time falls in converting hours into days and minutes, and combining the days, hours and minutes with the day and time of the insertion to determine the time of removal. Errors of a day become not only possible, but easy. One method to simplify keeping track of the days and half days begins by converting the insertion time into hours and fractions of hours instead of hours and minutes. For example, an insertion time on Tuesday of 2:43 p.m. becomes 14.72 (2 p.m. equals 14:00, and $43/60 = 0.72$). For a duration of 63.13 h, the removal time becomes $14.72 + 63.33 = 77.85$. Subtract 24 h blocks to mark the days, as

Wednesday	leaving 53.85
Thursday	leaving 29.85
Friday	leaving 5.85.

Thus, the removal time falls on Friday at the time of 5.85 or 05:51 a.m. ($0.85\ \text{h} \times 60\ \text{min h}^{-1} = 51\ \text{min}$). Not a particularly nice time, but that is what junior staff are for.

With the severe consequences of a significant error in calculating the removal time, a check of this calculation by a second person becomes extremely important. A computer program that calculates the time given an insertion time and date also serves as an invaluable resource for assuring the accuracy of this calculation, although if the computer-generated time serves as the primary calculation, it also requires verification.

Setting a personal count-down timer, often incorporated on a wrist watch, to alarm just before the calculated number of hours provides some protection against miscalculation of the removal time while converting to days. Additionally setting an alarm before the removal time serves to remind one to attend to the patient in the press of other commitments.

Once calculated, all parties need to know the removal time to assure that the necessary personnel attend at the specified time. The patient and the patient's family members should also be informed of the removal time, not only to know when radiation isolation will be lifted, but to remind the nursing staff if no one arrives for the removal.

4.1.3. During the treatment

Although radiation safety activities must continue through the treatment, the only real quality check while the therapy proceeds entails verifying the correct positioning of the applicator. LDR gynaecological applicators tend to move considerably in the patient under the best circumstances (King *et al* 1992). This movement, while built into the dosimetry protocols implicitly by basing doses on historical experience, probably accounts for unexpected recurrences and complications and uncertainties in the survival and cure data. However, little can be done to improve the situation without compromising patient comfort. Of real concern, though, is whether the applicator grossly remains in location. Many of the reported misadministrations in the United States stem from the applicator coming out of the patient or the sources falling out of the applicator (sometimes with the patient's assistance). Visual inspection of the applicator by a physician familiar with the case twice daily provides some measure of assurance that the applicator remains in the general treatment vicinity.

4.1.4. At the removal

The two important points at source removal are that all of the sources come out, and that they come out at approximately the correct time. Satisfying the second item requires only arriving on schedule. Delivering the correct dose to within 1% requires removing the sources within 36 min of the calculated time for a typical 60 h implant, although this window narrows to 12 min for a 20 h boost treatment.

Two actions usually assure the removal of all the sources. The first action is to count the sources during removal. If possible, it is good to note each source strength at this point just as during insertion in order to verify the correctness of the treatment. If identification requires disassembling the source holders, this verification can wait until returning the sources to storage. Following removal, with the sources shielded and out of the room, a survey of the patient, rubbish and linens provides a final margin against leaving a source in or near the patient (or staff).

4.2. APPLICATIONS INVOLVING TEMPORARY SOURCES ORDERED FOR THE TREATMENT

Many of the procedures for implants using sources ordered for the particular patient follow those described above for an implant using sources from a permanent inventory, and only the differences will be highlighted here. As a model, the discussion assumes the use of ^{192}Ir sources in plastic ribbons unless otherwise noted.

4.2.1. Source loading

Receiving the sources that the plan called for requires the successful and correct execution of three steps:

(1) transferring the planned source distribution into an order,
(2) transmitting the order to the vendor,
(3) assembling and labelling the source ribbons by the vendor.

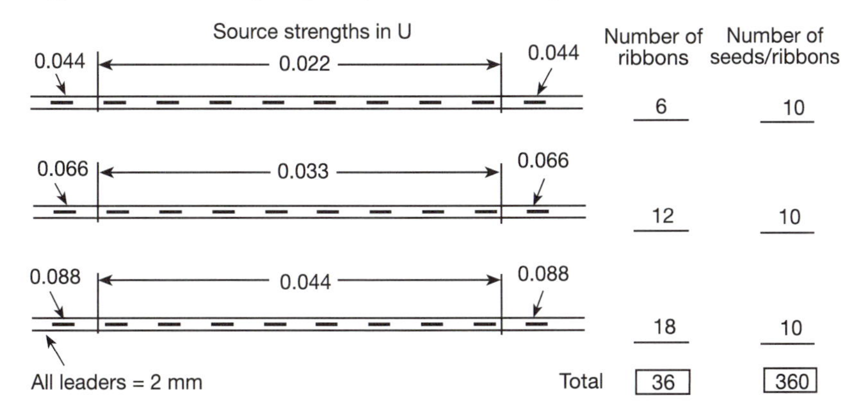

Figure 4.1. *Sample order diagram for a Syed–Neblett rectal template implant.*

The ease of transferring the source distribution from the plan to an order depends on the output of the treatment planning system. Some systems list the sources in the order entered. If the entry was from CT image-based planning, the order of entry may be quite random and the output very confusing. At the other extreme, the source entry may be specific to a needle track, and the output may list the sources in their positions along the track. In the latter case, the output almost serves as an order form. The complexity of the order, of course, depends on the implant's level of customization. An order for a uniformly loaded Syed–Neblett rectal template (with the hole pattern of six equally spaced holes in a circle of 1 cm radius, 12 holes in a circle of 2 cm radius and 18 holes in a circle of 3 cm radius) need only specify the source strength, the number of sources per ribbon and the spacing between and the number of such ribbons: a total of four numbers. The same implant, if optimized, would use different strength sources in the inner, middle and outer rings. The strengths of the sources along the needle would also vary, with at least the end sources being stronger than those between. In this case, the order includes at least three types of ribbon and different source strengths within each ribbon, and the order becomes less clear if presented as just a list of the parameters (although this order constitutes about the maximum complexity for such specification short of individually customized loading based on images). A figure clarifies the order, as in figure 4.1. For CT-based treatment plans, each needle assumes a unique configuration, and the order requires some sort of detailed specification. Figure 4.2 shows a sample order form.

Transmitting the order to a vendor can be a simple phone call for the uniformly loaded case, but for the more complex situations a fax or e-mail with a detailed order form becomes necessary. In general, without a written form, this step becomes a weak link in the process, prone to misunderstandings and errors. A transmittal document reduces the incidence of mistakes to extremely low levels.

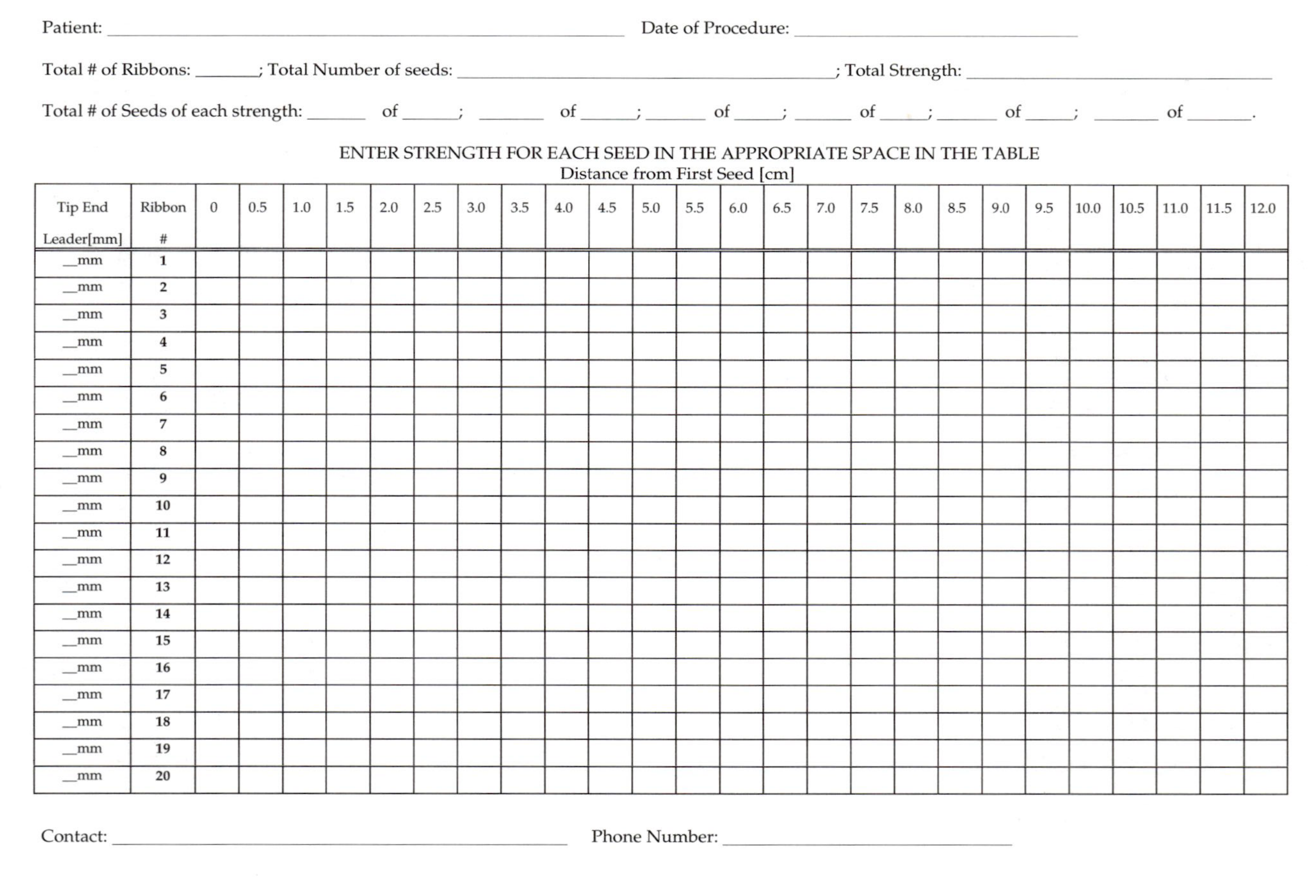

Patient: ____________ Date of Procedure: ____________

Total # of Ribbons: ______; Total Number of seeds: ____________; Total Strength: ____________

Total # of Seeds of each strength: ______ of ______; ______ of ______; ______ of ______; ______ of ______; ______ of ______; ______ of ______.

ENTER STRENGTH FOR EACH SEED IN THE APPROPRIATE SPACE IN THE TABLE

Distance from First Seed [cm]

Tip End Leader[mm]	Ribbon #	0	0.5	1.0	1.5	2.0	2.5	3.0	3.5	4.0	4.5	5.0	5.5	6.0	6.5	7.0	7.5	8.0	8.5	9.0	9.5	10.0	10.5	11.0	11.5	12.0
__mm	1																									
__mm	2																									
__mm	3																									
__mm	4																									
__mm	5																									
__mm	6																									
__mm	7																									
__mm	8																									
__mm	9																									
__mm	10																									
__mm	11																									
__mm	12																									
__mm	13																									
__mm	14																									
__mm	15																									
__mm	16																									
__mm	17																									
__mm	18																									
__mm	19																									
__mm	20																									

Contact: ____________ Phone Number: ____________

Figure 4.2. *Sample order form for ^{192}Ir sources in ribbons.*

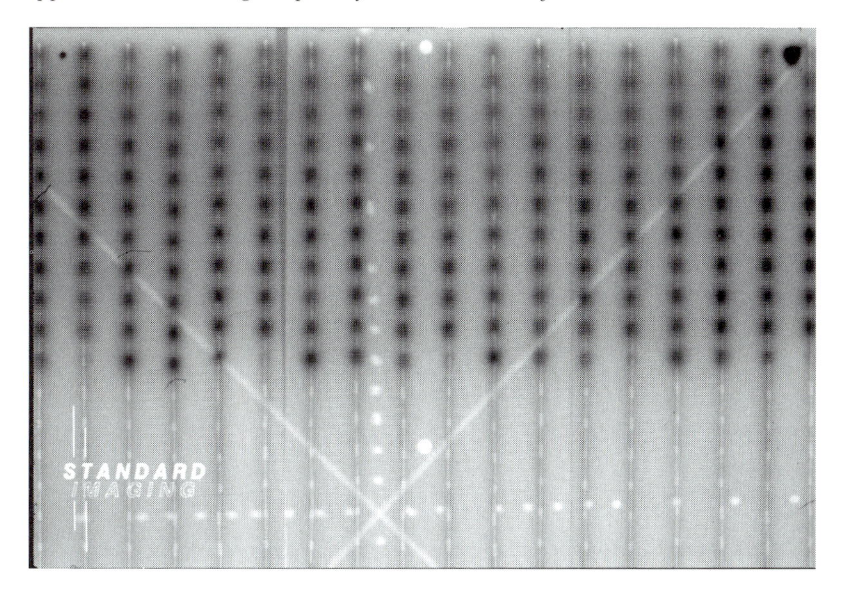

Figure 4.3. *Autoradiograph of* ^{192}*sources in catheters. Figure courtesy of Standard Imaging, Middleton, WI, USA.*

The last of the three steps lies entirely under the control of the vendor. The practitioner can only perform the first two steps as well as possible to assist the manufacturer. However, upon receipt, the institution assumes the responsibility to check the correct fulfilment of the order. The simplest check assures the correct number of sources in each ribbon. Particularly for customized implants, where the spacing between sources varies among the ribbons, a check of the source pattern provides an indication of the identity of some ribbons. While measuring the strength of a source is a straightforward procedure, checking the strength of individual sources in a train while still in a plastic ribbon poses considerable problems (as discussed in chapter 2). Making an autoradiograph of the sources in the ribbons as described in chapter 2 allows a visual check that the sources contain activity (there have been cases of nonactive 'duds' in a source train) and a differentiation between major source strength groups. Figure 4.3 shows an autoradiograph of a set of source ribbons.

Unfortunately, autoradiographs provide no quantitative information about source strength due to the contribution of radiation from other sources to the image corresponding to a given source. Short of following a deconvolution procedure such as that of Thomadsen *et al* (1997) as discussed in chapter 2, the next best check measures the strength of the ribbon as a whole. A measurement of the ribbon strength usually passes the ribbon through a looped holder in a calibrated well chamber, as shown in figure 4.4. Unfortunately, the calibration factor for sources in the loop often differs from that for single sources in the centre of the

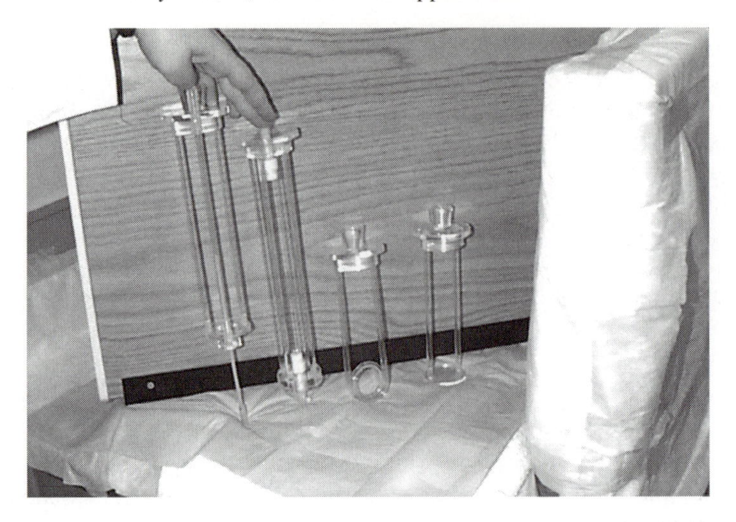

Figure 4.4. *Inserts for a well chamber. From left to right, inserts for measuring: tube-type sources and needles (with an adjustable height to centre the source), single seeds (with threads to hold sources in place without significant attenuation), seeds in ribbon or wire, liquid or bunches of seeds in vials.*

chamber. This difference in response comes from some of the sources in the middle of the loop lying closer to the chamber wall than the centre of the chamber, and other sources on the top and bottom falling above or below the centre. For uniformly loaded ribbons, simple correction factors depending on the number of sources in the ribbon convert the reading to the total source strength. However, for differentially loaded ribbons, the position of each individual seed determines its contribution to the total signal. For this case, the best check compares the expected reading based on the signal each seed should produce to the measured signal.

4.2.2. Calculation of the removal time

Several factors complicate the calculation of the removal time compared to that discussed in the previous section.

(a) *Correction for radioactive decay*. Because the sources used for customized temporary implants use radionuclides with relatively short half-lives, the calculations sometimes need to include corrections for radioactive decay *in situ*. Instead of the simple equation (3.2), equation (3.1) may apply. Table 3.8 provides guidance as to when equation (3.2) can be used.

(b) *Accounting for source insertion and removal*. Customized interstitial implants often consist of many needles or catheters. Pelvic templates, for example, frequently use more than 30 needles. Loading that many needles may take 15 to 30 min. While most catheter implants include fewer tracks, loading them often takes longer per track than needles. Part of the loading time for both techniques entails checking the ribbons for the correct number of seeds and loading pattern, if applicable. Removing ribbons from template needles takes little time, but each ribbon removed requires verification of the number of sources enclosed, and that check should be performed as each ribbon leaves the needle to facilitate corrective action should a source fail to be removed. In addition to counting the number of sources per ribbon, unloading catheter implants usually requires cutting through the catheter, but not the ribbon, a sometimes slow process. Thus, for any custom implant, the total loading and unloading time may exceed a half hour. While negligible for a 5 d implant, for a 20 h boost treatment this time becomes 2.5%. Were the case complicated or large, requiring half-hour loading and unloading, this time would amount to 5% of the total. Trying to account for this time presents the question of whether the treatment time begins as the first source goes into place, at some midpoint of the loading or when all the sources are in place delivering the planned dose rate. Ideally, the loading and unloading should form complementary procedures, each following the same order and taking the same time. Were that the case, the correct removal time would *begin* at the calculated duration after the *beginning* of the loading. Unfortunately, loading frequently takes longer, and may, for convenience, follow a very different order from unloading. Thus, outside the ideal situation, there is no 'correct' time from which to begin unloading. As a compromise, many practitioners use the middle of the loading process as the in-time, and begin unloading at the calculated duration following that mean time. This usually produces a slightly long true net treatment duration, but by a small enough margin to be ignored.

4.2.3. During the treatment

The most likely time to misplace a source used for intracavitary treatments using sources from a permanent inventory occurs during loading. The closed applicators used with that procedure render source loss unlikely through the remainder of the treatment duration. The temporary interstitial implants, on the other hand, most often lose sources during the relatively long quiescent period between loading and unloading. Three situations reported frequently include:

- source ribbon tails snagged by gauze covering the butts of template needles while changed by a nurse and discarded in the rubbish;
- tongue or flood of mouth sources coming out during a meal and removed with the food tray;
- a nurse or the patient 'adjusting' an endobronchial catheter and pulling out the source.

Educating the nursing staff to recognize the sources and source ribbons and check for their placement stands as the first line to preventing source loss, and erroneous treatments for patients. In addition, the physician should check the integrity of the implant at least twice daily through the duration. Welding each source ribbon to its catheter helps prevent the loss of ribbons from catheter-based implants; however, should a button come loose, the catheter-and-ribbon unit still may leave the proper site.

4.2.4. *At the removal*

Removal becomes a critical time for source accounting with temporary interstitial implants. Only friction holds the sources in the ribbons. On contact with fluids, the ribbons act like capillaries and pull the fluid into the lumen. Once wet, particularly with body fluids, the sources slide along the ribbons easily and can leave the ribbon. Fluids in the ribbons become a major problem with open-ended needles—tools that should be avoided with any implant if at all possible. A source out of the ribbon and in a needle can fall out during needle removal, and lodge in the patient's skin folds or other sites, or leave the room in the linen or rubbish. With catheter-based implants, the source could fall from the ribbon into the catheter, and during removal of the catheter, slide from the catheter into the patient's tissue along the catheter track, requiring surgical removal. To avoid such unpleasant occurrences, the number of sources in each ribbon should be counted as removed from the needle or catheter, and compared with the loading diagram.

4.3. APPLICATIONS INVOLVING SOURCES ORDERED FOR A PERMANENT IMPLANT

A permanent implant requires fastidious quality control because there is no second chance: once implanted the sources stay in place permanently. The occasion for checks only happens at the time of implantation, for obvious reasons.

4.3.1. *Source strength check*

Permanent implants use either loose sources or sources in resorbable suture material. Almost all permanent implants use uniform-strength sources. The check of the source strength differs for each. The loose sources go readily into a well-type chamber, as discussed in the section of chapter 2 on source strength assay. The report of Task Group 40 of the American Association of Physicists in Medicine (AAPM TG40 1994) recommends assay of a random 10% sample of the sources intended for implantation. The check of the assay of the seeds actually provides more than a verification of a manufacturer's ability to calibrate the sources; the assay also forces one more opportunity to check that the strength sources received actually correspond to the strength designated in the plan, and that no mistake occurred between the placing and filling of the order.

A sampling as recommended takes little time and provides a 10% chance of detecting a dud among the sources. For a 100-seed implant, a single dud only slightly perturbs the dose distribution, and that only locally. Yet if that local perturbation falls at a critical location where few other seeds neighbour, the consequences may be treatment failure. Multiple duds increase the deficiencies of the treatment while only increasing the probability of detection slightly.

Sources in resorbable suture come to the user sterilized in a shielded tube. As discussed in chapter 2, this makes verification of the source strength before loading difficult at best.

4.3.2. Source position checks

Checks that the implantation will place the sources in the desired position depend on the delivery route. The two most common delivery techniques for permanent implants use preloaded needles or a source-deposition device.

4.3.2.1. Needles preloaded with sources. For the loading of the needles, each needle lumen is conceptually divided into positions approximately the length of the source. Each position contains either a source or a source-length spacer (made of stiffened resorbable suture). Since all sources contain the same strength, customization of the dose distribution for permanent implants relies on differential loading of source locations. To achieve the planned dose distribution requires the correct loading of the sources into the positions in the needles and the proper placement of the needles in the patient. Many of the devices designed to facilitate loading the needles allow a check of the order of the train of sources and spacers just prior to insertion into the needle. Figures 4.5(b)–(d) show such devices. However, the device in figure 4.5(a) loads sources or spacers one at a time into the needles, and opacity prevents any verification of the order of the contents of the needle.

As a final check on needle contents, the stylettes for the needles are marked in 0.5 cm bands, approximately the size of a source. Counting the number of bands above the top of the needles gives the approximate number of sources and spacers in the needle. Two factors prevent the stylette marks from disclosing the number of sources and spacers precisely. The first stems from the material closing the tip of the needle. This material, either bone wax or anusol, forms plugs in the tip of varying lengths and is hard to control accurately. The second cause comes from an attempt to add accuracy to the position of the sources. Most sources used for permanent implants currently measure 4.5 mm in length. Since steps during planning locate sources in planes separated by 5 mm, the manufacturers of the spacers make them 5.5 mm long, so that trains of alternating sources and spacers keep the source centres at centimetre intervals. This odd length for either spacers or sources causes placement problems for long strings of either as opposed to alternating the two. Optimizing implants seldom calls for strict alternation because of the irregular shape of targets. With customized loading based on imaging studies (such as ultrasound or CT), needles often contain several sources or spacers in a row. The length of such trains deviates slightly from integer multiples of 0.5 cm. From this cause alone, a needle with five sources in a row indicates 2.5 mm too short

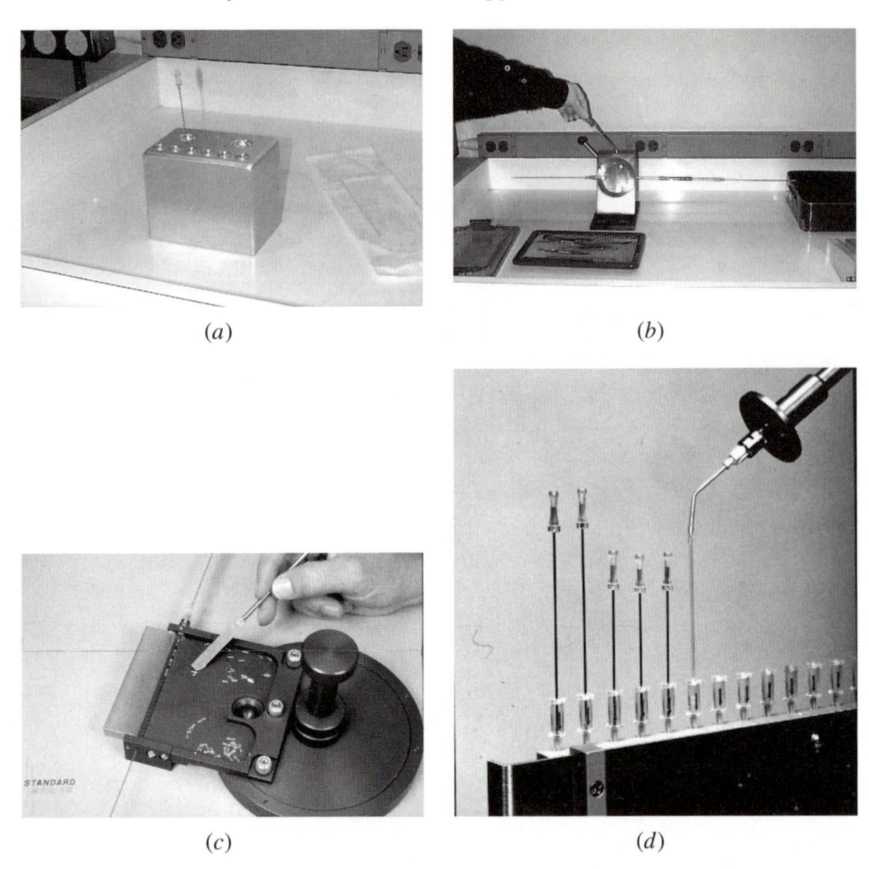

Figure 4.5. *Tools to assist loading needles with seed-type sources for permanent implantation. (a) A simple shielded needle holder for dropping sources or spacers into needles one at a time. (b) A device where seeds load into a channel from the top. The clear window allows verification of the loading. When the train is complete, the channel rotates to the horizontal, and the train passes into the needle. (c) A device where seed trains are assembled in a groove, and then slid into the needle. (d) and (e) A vacuum system where the seeds and sources are sucked into a glass holding tube in the reverse order, and following verification, dropped into the needle by releasing the vacuum. (Figures (c) and (d), courtesy of Standard Imaging, Middleton, WI; (e), courtesy of Medical Radiation Devices, Dothan, AL.)*

on the bands, and that with five spacers 2.5 mm too long. Sources in resorbable suture only come with centimetre spacing between source centres, eliminating the uncertainty in the relative positions, but limiting optimization.

From loading, the needles often go into a shielded box with holes for the needles in the same pattern as the implant template (see figure 4.6). Mistakes

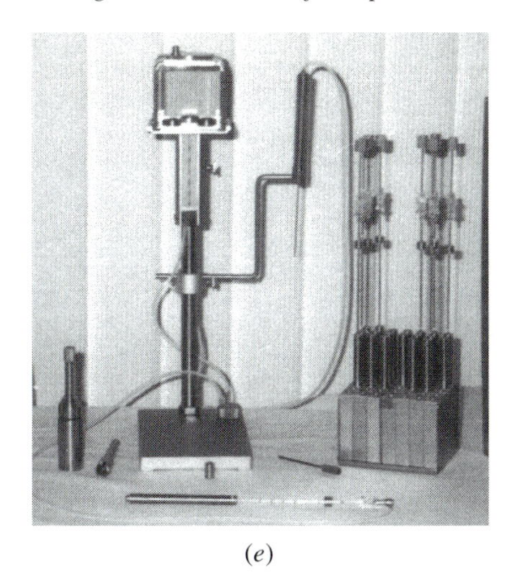

(*e*)

Figure 4.5. *(Continued)*

in placing the needles into the proper hole easily translate into implanting the sources in the wrong location in the patient. Having two persons at the needle preparation, one performing the loadings and the other checking the source/spacer pattern and the position of the needle in the box, provides a considerable degree of protection against errors. Placing pieces of tape over unused holes before starting the loading process also prevents the common error of simply putting a needle in a hole that should remain empty. Working from top to bottom and left to right (or vice versa or in any fixed pattern) and leaving only the intended holes clear to accept a needle, a misplaced needle usually becomes obvious with the placement of the next needle, or at least at the end of a row. Since the loading procedure often is performed under sterile conditions, if sterile tape is not available, spare stylettes can be used to fill the unused holes. Without a second person watching during loading, the sole person needs to methodically work through the loading patterns. Highlighting each needle on the loading diagram as the needle is completed and placed in the box helps prevent skipping a needle or repeating the pattern erroneously. One method to ensure the proper loading of the needles in the proper location would have the rows of needles in the shielded box consist of slide-out trays. Each tray could be placed on a sheet of film sterilely and radiographed. The resultant image displays the loading pattern (particularly if the trays contain lead markers showing the template coordinate locations) and forms a record of verification.

4.3.2.2. Source deposition device. Examples of source deposition devices include the Mick applicator (figure 4.7(a)) and the Royal Marsden implant gun (figure 4.7(b)). Each of these devices delivers a single source at the tip of the

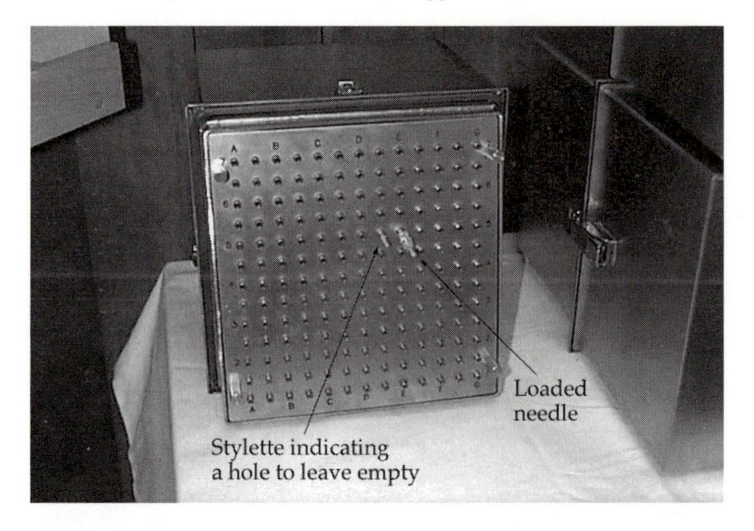

Figure 4.6. *A needle box for holding the implant needles between loading and the procedure. The grid pattern for the holes in the box assists the physician in placing the needle in the correct hole of the template. The figure shows a loaded needle in the box and an empty stylette filling a hole not used in the plan to avoid accidentally placing a needle at that location.*

needles with an implantation cycle. The Mick applicator design includes several features to assist the operator to place seeds at specified intervals along a needle track. The ring around the needle sets against the patient's skin or the template block (depending on the model) once the tip of the needle reaches the target depth. Alternatively, the ring can be set to allow the needle to penetrate only to the desired depth. The operator pulls the stylette fully out, allowing one seed to move from the spring-fed magazine into the needle axis. Plunging the stylette fully in delivers the source just out the tip of the needle. Holding the ring on the skin (or block) and pulling on the body of the applicator moves the tip of the needle outward along the track. A spring-fed ball bearing drops into a depression in the frame each time the applicator moves 0.5 cm. After moving the distance between sources, withdrawing and plunging the stylette dispatches another source. Care should be taken after placing a source not to withdraw the stylette until after moving the body of the applicator and the needle. To do so can draw the freshly implanted source back into the needle. When this happens, that source is pushed out of the needle by the next source, resulting in two seeds at the same location. Sucking the seeds back into the needle can happen several times along a track, resulting in several sources all implanted at the last source's position.

With moving the applicator and taking care to deliver one source at a time, it frequently becomes difficult for the operator to keep track of the pattern for each needle as the loading progresses. A second person in the operating room following

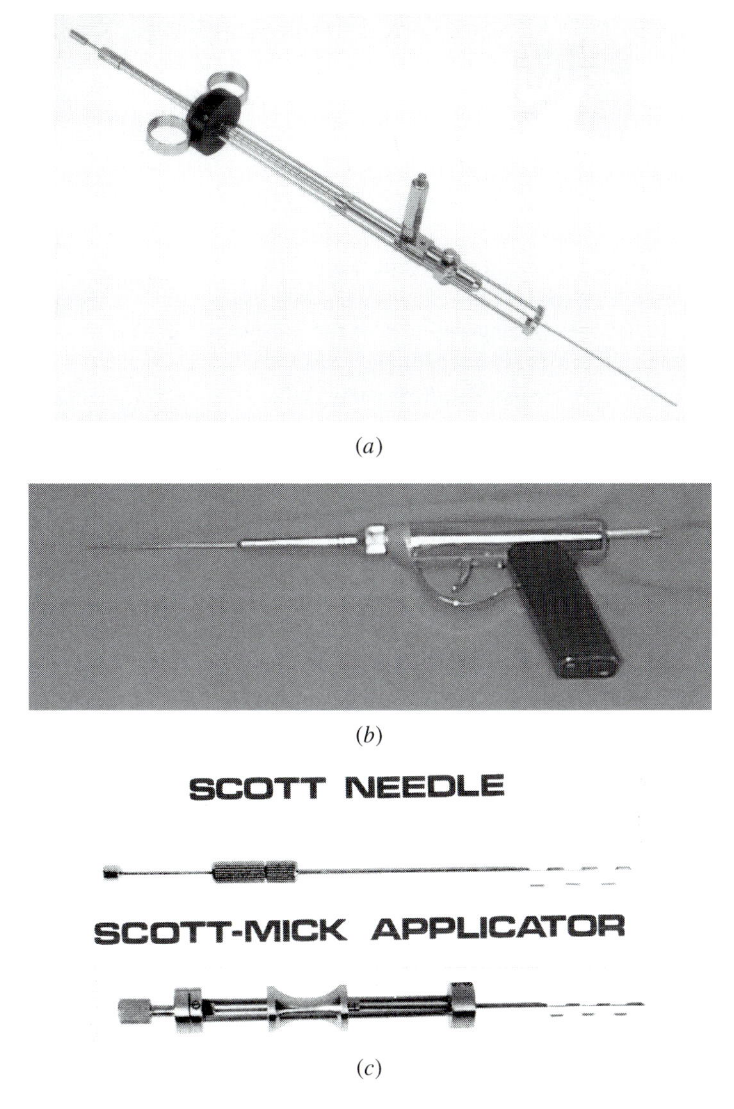

(*a*)

(*b*)

(*c*)

Figure 4.7. *Seed delivery devices. (a) The Mick applicator. (b) The Royal Marsden gun. (c) The Scott applicator. Figures (a) and (c) courtesy of Mick RadioNuclear Instruments, Bronx, NY.*

the loading with the diagram guiding the operator becomes almost essential for all but very small implants with a single loading pattern for all needles.

The Marsden gun delivers one seed each time the operator pulls the trigger. This device simply implants multiple seeds, and contains no mechanisms to assist in positioning those sources. Some of the guns tend to jam if operated too rapidly, and this can result in misplaced sources. Delivering precision source patterns becomes difficult with this device, particularly without a second person to assist in measuring needle retraction and accounting for sources as they are placed.

4.3.2.3. Hybrid applicators. Some applicators bridge the difference between preloaded needles and the single-source deposition devices. One example is the Scott applicator (figure 4.7(c)). This applicator places sources in precut positions in a loading needle. A sheath covers the needle during insertion. Withdrawing or rotating the sheath exposes the sources, and upon rotating the needle the sources stay in the tissue. Depending on whether the needles are loaded before coming to the operating room or in the operating room just prior to insertion, such applicators follow the nature of the preloaded needles or source deposition devices, respectively.

4.3.2.4. Positioning without a template. The foregoing discussion assumed a template-guided placement of the needles. Some applications prevent the use of templates. For freehand applications, accurate placement of sources becomes incredibly challenging. If possible, the physician should try to insert all the needles in their places and verify the needle positions radiographically or by ultrasound before delivering the first source. Often, however, the same limitations that prevent the use of a template prohibit placing all the needles simultaneously. If possible, at least the needle pattern should be marked on the patient's skin with ink, and frequent fluoroscopic evaluation of the implant progress performed.

4.4. IMPLANTS USING RADIUM OR CAESIUM NEEDLES

Implants using radium or caesium needles find little application nowadays since temporary sources of ^{192}Ir are available almost in any part of the world. The radium or caesium needles limit the depth of treatment coverage, pose the danger of rupture during implantation, and expose the hospital staff to radiation needlessly. Within these limitations, caesium and radium needles can provide effective therapy. None the less, implants using such needles also should follow guidelines to assure accurate treatment. As with other interstitial implants, templates assist in correct needle placement. Unlike the templates used with afterloading needles, because of the short physical lengths available for these 'hot' needles, the templates must be thin if they remain in place during treatment, or removable if they are thick enough to guide the needle angle as well as the insertion position. Figure 4.8 shows a two-part template that guides the needles into place, and then separates for removal allowing suturing of the needles to the patient.

Most radium or caesium needles come in two strengths (full or half strength), usually identified by colour (gold or nickel plating on caesium, gold or platinum for radium). These strengths require review just prior to implantation just as with

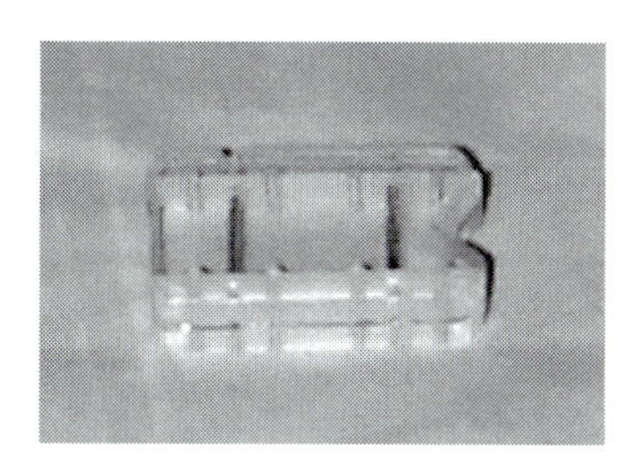

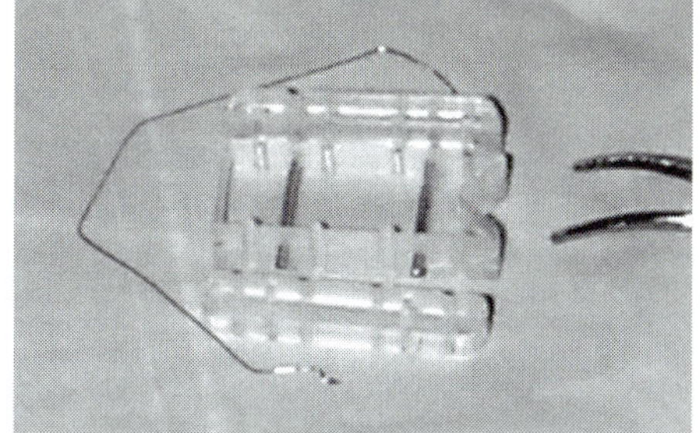

Figure 4.8. *Template for use with caesium needles. The needles pass through holes in the plastic. Unwrapping the wire around the template allows removal of the template from around the needles so the needles can be sutured in place.*

the intracavitary sources. The physical lengths of the needles most often come in sets of 2.7, 4.2 and 5.7 cm, which most physicians (and even physicists) can distinguish readily through observation. However, more complete sets include 3.2 and 5.2 cm sources, which look little different from some of the needles in the basic set. With the more complete set, the length also should be measured just prior to insertion.

Since no template holds the needles in place during the treatment, sutures through the eyelets secure each needle individually.

4.5. SUMMARY

Loading sources into a treatment appliance forms one of the most critical junctures for the entry of errors into a brachytherapy treatment. The key for any type of implant is to ensure that the proper sources arrive at the correct locations. For temporary implants this may be checking a colour code or verifying a source train configuration. For permanent implants with sources of a single strength, this check mostly entails verifying the strength of seeds and delivery of seeds in the predetermined pattern. Correctly implanting source trains ordered for an individual patient requires clear and accurate communication to the supplier of the source train configuration for the treatment planning system.

Basic to accurate treatments, the removal time calculation for temporary implants needs verification by a second person, and a reminder, such as an alarm, for the person removing the sources.

During temporary-implant treatments, the physician should check the sources in the appliance about twice daily to assure no major changes have occurred.

Ensure that no temporary sources remain in a patient by counting the sources during removal, and then, with the sources shielded or out of the room, surveying the patient and the patient's room.

Table 4.2. *Quality management steps at the time of low dose rate brachytherapy treatment applications.*

Step
Ordering sources
Verify transcription from planning computer to order form
Verify strength of sources on receipt
At source loading
Temporary applications
Identify source strength and source position in appliance
Assure sources are secure in appliance
Calculate removal time, and verify with a second person
Set a timer to alarm before removal time
Permanent implants
Verify needle-loading pattern
Check seed deposition during the procedure
During treatment
Check on implant twice daily
At removal
Begin removal at the correct time
Count all sources upon removal
Shield sources and survey patient and room
Recount sources on return to the storage facility

Table 4.2 summarizes the quality management steps to take at the time of an LDR brachytherapy procedure.

CHAPTER 5

QUALITY MANAGEMENT FOR HIGH DOSE RATE UNITS

Probably more than with any other form of brachytherapy, high dose rate (HDR) treatments present a high potential for accidents and errors for two reasons:

(1) the planning and treatment require intensive interactions between human operators and sophisticated machines and computers, where small errors can produce large effects, and
(2) planning and execution of the treatment takes place over a very short time, with each step happening quickly, so errors can easily be executed before detection.

Conventional, low dose rate brachytherapy usually delivers the treatment over a period of days, which allows time to review the treatment plan and parameters, making adjustment as necessary mid-treatment. Most external-beam radiotherapy consists of many fractions, and errors in early treatments can be corrected in later sessions. HDR brachytherapy offers neither the latitude of time nor many fractions. Only a well developed and disciplined quality assurance programme stands between a safe practice and disaster.

This chapter addresses quality assurance for the HDR treatment unit, but, as such, only addresses part, the most obvious part, of the QM necessary for consistently satisfactory treatments. The next chapter covers QM for the treatment plan, which forms the rest of the programme.

All HDR units use internal interlocks to prevent initiation of a source excursion or to automatically initiate a source retraction if the unit detects an unsafe, inappropriate or contradictory condition. Much of the testing involves challenging these interlocks to assure their proper operation.

As with the previous chapters on hardware, the evaluations fall into categories of 'initial' and 'periodic checks'. In this case, however, the initial checks (with a few exceptions) occur with each source change (approximately every three months), while the periodic checks fall at the beginning of each day the treatment unit sees use. While, for the most part, safety considerations fall outside the considerations of this text, with HDR units, the safety checks and interlocks often serve to prevent erroneous treatments, and so play a significant part in quality assurance. Readers should refer to applicable regulatory guidelines for their venues that may dictate tests and periodicity that must be followed and may differ from the following suggestions. Where appropriate, recommendation for test frequency follows that of the report of Task Group 59 of the American Association of Physicists in Medicine (AAPM TG59 1998). Another excellent reference is Williamson *et al* (1994). Much of current quality management for high dose rate brachytherapy results from early work by Ezzell (1990, 1991), Chenery *et al* (1985), Flynn (1990), Grigsby (1989), Jones (1990) and Meigooni *et al* (1992).

5.1. PERIODIC EVALUATIONS (AT THE BEGINNING OF EACH TREATMENT DAY)

The list below specifies a set of tests that evaluate the important systems of an HDR treatment unit. A discussion of techniques for executing the tests follows.

(I) Safety checks:

(a) Check for proper operation of the audio and video communication between the treatment room and the control console.

(b) Check the interlocks prevent source exposure if:

- (1) no complete applicator is attached (including transfer tubes, adapters and patient appliances, as appropriate),
- (2) the applicators or transfer tubes are not locked in place in the indexer,
- (3) the path through the distal-most dwell position is obstructed, or the catheter has too tight a curvature for the source cable to negotiate,
- (4) no complete programme has been entered or
- (5) the door to the treatment room is open,

and retract the source:

- (6) if the door to the room opens,
- (7) if the 'treatment interrupt' button is pressed,
- (8) if the 'emergency off' button is pressed or
- (9) when the timer reaches the end of its setting.

(c) Check proper operation of the room radiation monitor:

- (1) that the monitor makes a gentle sound when exposed to the radiation background in the room with the source unshielded,
- (2) that the monitor, when activated, triggers warning lights visible by the room door and immediately upon entering the room.

(d) Check the proper operation of the hand-held Geiger counter.

(II) Dosimetry checks:

(a) Check the positional accuracy for the source movement.
(b) Check for consistency of the source strength.
(c) Check the consistency of the exposure timer and transit time.
(d) Check the linearity of timer settings.
(e) Check that the treatment unit contains the correct value for the source strength.
(f) Check that the controlling computer reflects the correct date and time.
(g) Check that the controlling computer correctly accepts a treatment program from the normal transfer medium.

5.1.1. Safety checks

While the list of safety check seems long, the evaluation need not consume a great deal of time. At our facility, the entire morning routine takes about 10 to 15 min, starting from uncovering the unit. Most of the items could be tested in numerous different manners, but only one set of techniques will be discussed here. Individual units may differ in the exact methods. The procedure in the list assumes the successful completion of all of the items going before. Failure of some of the items requires changes in the procedures for subsequent tests. Failure of *any* item requires evaluation by the physicist of the appropriateness of continuing with patient treatments in light of the particular failure.

Communication equipment. See that the television and intercom systems function. PASS: television images of the room are visible and sounds in the room can be heard; FAILURE: either communication device fails to operate. (Note that there may be no sounds to hear until later in the procedure when the unit begins to function. Audio verification can wait until that time.)

Applicator attachment. Program the unit to send the source to a location in each of the channels without attaching any catheters or transfer tubes to the unit, and set to have the check cable check each channel for patency. Some models require special settings for this test (such as the Nucletron's *special mode*); for others this test can only be performed on the first channel, which, after failing, will wait for a catheter to be attached to that channel before checking the second channel. For such units, this test proves too time consuming, and moves to a test performed monthly. PASS: the unit refuses to send the source out any of the channels; FAILURE: the check cable and/or source exits one or more of the channels.

Catheter attachment lock. Attach catheters to each of the channels but do not lock them in place, and try to initiate source movement. PASS: the check cable refuses to pass into any of the catheters; FAILURE: the check cable and/or source exits one or more of the channels.

Pathway patency. Remove all the catheters from the channels, and attach and lock to one channel a special catheter that has been curled into a loop with a radius of curvature too small for the source to negotiate near the end. Program the source to travel to the end of the catheter, and initiate a run. The check cable should detect the loop as a problem for the source. PASS: after the check cable run, the unit refuses to send out the source; FAILURE: after the check cable run, the unit sends the source into the catheter.

Program completion. Program the unit with some missing information, such as a single dwell position but with no time. PASS: the unit will not initiate source movement; FAILURE: the source passes into the catheter.

Door interlock. Program the source to a position short of the loop with about a 30 s dwell time. With the door to the room open, try to initiate a run. For some units in some conditions, the check cable may move through the catheter. PASS: the unit refuses to advance the source into the catheter; FAILURE: the unit sends the source into the catheter.

Warning lamps. Close the door, and initiate the source run. Observe the warning lamps by the door. Most units have a set of two lamps, one connected to the treatment unit that triggers when the source leaves its shielded housing, and the other that lights when the signal on the radiation detector in the room exceeds its trip level. PASS: the lamps come on when the source passes into the catheter; FAILURE: one or both of the lamps fail(s) to light.

Room monitor operation 1. Listen through the intercom for the sound of the room monitor. PASS: the monitor produces an audible sound; FAILURE: no sound can be heard.

Room monitor operation 2. Open the door to the room and observe the visual indicators on the room monitor. The room design should provide protection to a person in the doorway until the source retracts. PASS: the indicator lamp is lit; FAILURE: the indicator provides no signal that radiation is present.

Hand-held monitor. Immediately on opening the door during the previous test, hold the hand-held monitor in the doorway and see whether it indicates the presence of radiation. PASS: the hand-held monitor indicates a radiation reading; FAILURE: the monitor fails to respond. (Note: paralysable Geiger counters pose a serious hazard with HDR units. Such a meter responds to low levels of radiation, but exhibits decreasing to vanishing readings in high exposure fields due to overlapping dead time. The hand-held monitor must be tested during its commissioning by exposure to high level fields. As an easy way to test for this problem, the Geiger counter can be placed in the treatment room near the unit, and watched on the television monitor during a source run. The hand-held monitor should indicate a full-scale reading. Scintillation-based detectors usually saturate at levels of radiation exposure too low for use in the environment of a high dose rate unit.)

Door interrupt. During the previous two tests, the unit should have been retracting the source beginning from the opening of the door. PASS: the unit retracts the source; FAILURE: the exposure continues with the door open, the unit delays significantly (more than a second) initiating a source retraction, or the retraction takes longer than 8 s to return to its shielded location.

Treatment interrupt. Close the door and reinitiate the exposure. Once the source reaches the dwell position, press the 'treatment interrupt' button. PASS: the unit retracts the source; FAILURE: the source remains out of the shielded housing.

Emergency stop. Restart the unit. Once the source reaches the dwell position, press the 'emergency off' button. PASS: the unit retracts the source; FAILURE: the source remains out of the shielded housing.

Timer termination. Reinitiate the exposure and let it continue until the elapsed duration equals the time set on the timer. PASS: the unit stops the exposure and retracts the source; FAILURE: the exposure continues past the time for termination. (Note: the actual exposure continues slightly beyond the time set since at the termination time the unit begins retraction of the source. Any location continues to receive radiation as the source moves back through the catheter to the housing.)

Paper supply. Check that the supply of paper and ink (or equivalent) will last through the treatments. Some units will initiate a source retraction if they detect the inability to record the progress of the treatment. PASS: supply is adequate to allow treatment; FAILURE: the supply may not last through the treatment.

5.1.2. Dosimetry checks

The dosimetry checks also form safety checks, for failure of any of the dosimetry processes produces an unsafe condition for the patient. The thrust of these checks focuses on the delivery of the correct dose to the proper location.

5.1.2.1. Source positioning accuracy. Proper treatment requires that the source occupies the position along the catheter corresponding to that used in the treatment plan. This requirement leads to two further criteria: that the source goes to the correct location for the dwell position programmed, and that the positions used in the treatment plan correspond to the dwell positions. The dwell positions used in the treatment plan often come from radiographic images made with markers in the catheters indicating specified dwell positions. For example, dwell positions

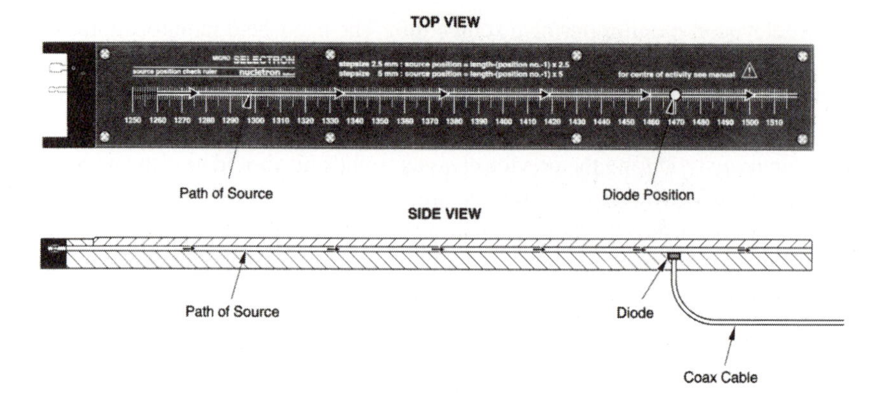

Figure 5.1. *A 'ruler' from Nucletron used to measure the length the tip of the source cable extends. The ruler shown in this figure contains a user's modification consisting of an additional diode used to quantify source position (Speiser and Hicks 1995). Figure courtesy of Nucletron-Oldelft, BV, Veenendaal. The Netherlands.*

often fall every 2.5 cm along the catheter. The markers (often called 'dummies' in the vernacular) may consist of a long wire with nubs ('indicators') attached at centimetre intervals, indicating every fourth dwell position. The marker wire usually has a flange on the back that abuts the rim of the catheter when seated in the correct location if the unit determines length that way, otherwise the wire seats at the tip end of the catheter. With the markers thus situated, the dwell positions assume definite, as opposed to simply relative, positions. To direct the source to correct locations corresponding to each dwell position, the source controller requires the distance along the catheter corresponding to the first dwell position. The distance may refer to the length from some part of the unit (such as the front face, the point of catheter insertion or a microswitch that tells the unit when the source enters the catheter), or it may be from some fictitious point (similar in concept to the effective source for electron beams). Establishing or verifying that distance that positions the source at the same location as the first dwell position indicated by the marker becomes an important part of the morning quality assurance.

The distance to the first dwell position can be established many ways. The first uses a specially designed ruler that replaces a catheter for the source track (figure 5.1), marked directly with the 'distance'. Sliding the dummy wire into the ruler and recording the reading for the first dwell position establishes a baseline distance. Attaching the ruler to the unit and focusing the television camera on the scale allows the position of the tip of the source wire to be measured during an exposure. Knowing the geometry of the source construction, the length from the tip to the centre of the source material can be *assumed*, and distance calculated for the tip of the source wire that places the centre of the source material coincident with the centre of the first dummy marker. The assumption necessary for this determination places an important uncertainty on the outcome. A poorly constructed source, or a change in source design not known to the operator, could result in erroneous source placement during treatment. As an additional problem executing this test, visualizing the ruler's scale adequately to distinguish the reading requires an extremely good television system.

The next method utilizes autoradiographs. Tape a clear, or at least translucent, catheter to a piece of paper-jacketed film (for example Kodak™ XV-2 'ready pac'™). Insert the dummy wire. With a pin, prick holes through the jacket and film on both sides of the catheter at the position of the first dwell position. Poking holes in both sides gives some indication of the uncertainty in marking the position. Do the same for the dwell positions 5 cm and 10 cm from the first, and darken the room. Program the source to stop at the marked positions using the best known value for the distance, with dwell times of about 2 s each for a source with a strength of 0.04 Gy m^2 h^{-1}. The times for other source strengths vary in inverse proportion. Execute the run and process the film. The resultant image looks like figure 5.2(a). The centroid of the dark 'blot' indicates the effective centre of the source, and should fall on the line between the pinpricks. The difference between the centroid and the pinprick line gives the length to be added to or subtracted from the distance to have the source coincide with the first dwell position indicated by the markers. The difference between the centroids for the other two dwell positions and their respective lines should be identical with the difference for the first dwell position. Changing differences means that the step size executed by the unit differs from that on the dummy marker. Use of this method requires several caveats.

- Darkening the room and processing the film quickly without significant exposure to light keeps the part of the blot produced by light to a minimum. Allowing the film to sit in light for more than a couple of seconds and/or using too short exposure times results in losing the radiation-induced 'blot' in the light-induced blot. Such a film will always indicate that the source stopped exactly between the pinpricks since the darkness comes from the light streaming through those holes. Figure 5.2(b) shows a film with the useful information masked by the light leak.
- The dwell times programmed for the three positions should be long enough so that the radiation induces a blot approximately twice the diameter of the light-induced blot. On each daily film, a pinprick off to the side, away from the source track, provides an indication of the light-induced signal. The dwell times should also remain short enough that the blots stay less than 3 cm in diameter. Larger blots add to the uncertainty of the centroid position, and can start to interfere with the edges of adjacent blots.
- The centroids of the blots should be determined using a ruler, finding the centre from where the optical density appears the same above and below the pinprick line. Just picking a centroid by eye increases the uncertainty to

about equal to the acceptable limits of the test, and tends to place the centre nearer to the pinprick line.

- A faster film than Kodak XV-2, such as Kodak XTL-2, gives unreliable results. With increased film speed, the light-induced artifact becomes larger, and the dwell times to keep from simply blackening the film become so short that the source-produced signal is lost in the artifact. With the very short dwell times, the difference between the darkening produced with the source dwells in the position and that produced with the source in transit becomes slight, increasing the uncertainty, and moving any centroid for the first position toward the unit (because the source approaches from and retreats in that direction, never spending any time on the opposite side).

For frequent testing, an alternative to taping a catheter to a film and poking holes fixes the catheter permanently to an acrylic sheet. The locations of the dwell positions of interest are marked with lead wires, also glued to the sheet. While not necessary, milling a slot for the catheter in the sheet makes for a more convenient device. Figure 5.3 shows the final product. This device rests on the paper-jacketed film during the same exposure programme as described above. The shadow of the lead wires should bisect the resulting darkened blots. Placing the last wire 1 mm off from the proper position gives an indication on the film of how an error in the first two positions would look. A device of this nature requires frequent checks that the wires correctly indicate the dwell positions (by inserting the dummy wire in the catheter), because, with daily use, the catheters tend to stretch.

After determining the proper length to send the source to the same location as the dummy-indicated first dwell position, the criterion for repeating the test with the correct length becomes, PASS: the position matches the dummy position to within 1 mm; FAILURE: the difference exceeds 1 mm.

Once the correct distance has been determined, a modification for a well-type ionization chamber allows more convenient testing in the future. The modification consists of lead cylinder inserts that fill the volume of the well except for two apertures, or slits separating the cylinders (DeWerd *et al* 1995). Figure 5.4 shows the configuration. One aperture coincides with the most sensitive position along the chamber, while the other falls about 3 to 4 cm away. The current signal measured as the source moves through the chamber looks like the curve in figure 5.5. The point marked 'A' falls in the middle of the slit coinciding with the maximal response of the chamber. At this position, the signal varies little with small changes of position. The points marked 'B', 'C' and 'D' fall on the edges of the slits where the change in signal as a result of a change in position assumes the largest values. Thus, while point A remains fairly immune to small positional changes, the other three points become eminently sensitive. Dividing the readings at points B, C and D by that at point A gives ratios that become independent of variations in the source strength as the source decays. A 0.5 mm variation in the source position produces a 10% change in this ratio. Thus, sending the source to these points during the daily QA routine provides a measure of the source positional accuracy. A table can be

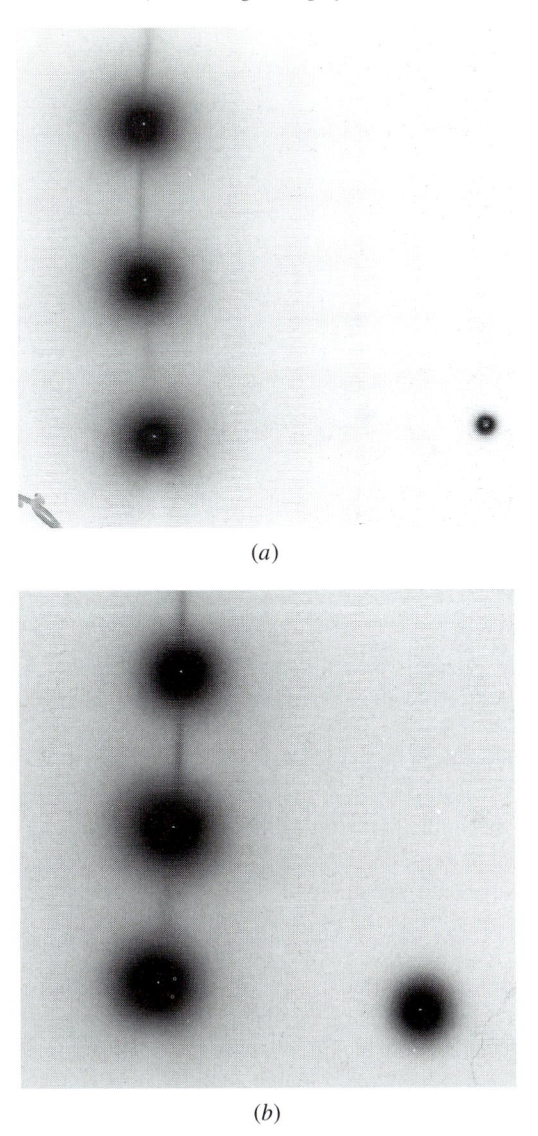

(*a*)

(*b*)

Figure 5.2. *Images for assessment of source positioning. (a) shows correct centring of the dark blotches made by the source with the pinprick holes. The small blotch to the right indicates the contribution of light leaking through the pinprick. (b) shows a film that was left in the light too long, and the blotch to the right approximates the size of the darkenings at the test sites.*

established that specifies the length to send the source so it coincides with the first marker position, based on the ratios measured. In making the table, a ratio that indicates that the source fell short of the standard length implies that the correct

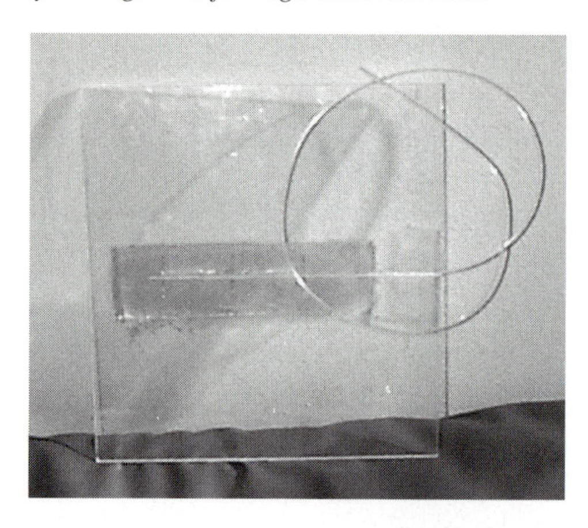

Figure 5.3. *Jig for source position checks. The jig sits on a piece of film and the source moves through the catheter (horizontal in the figure), stopping over each of the lead wires. The wires cast shadows on the film, which, if centred in the darkening from the source, indicate the correct value for the length.*

length to send the source is now longer than normal by how far the length was short. (Measured short distances require longer distance settings.) The source pathway into the chamber uses a needle instead of a catheter to avoid the problems of catheter stretching and kinking, and gives more consistent results.

Hicks and Ezzell (1995) and Speiser and Hicks (1995) describe a position-checking device of similar function to the differentially shielded well chamber. Their device utilizes a diode detector inserted into a source-positioning ruler, about 4 mm from the source path (see figure 5.1). The proximity to the source path makes the response very sensitive to the position of the source, with a maximum slope of approximately 16% of the peak reading per millimetre.

Ideally, each channel to find use on a given day should have the length tested that morning. For practices treating intracavitary gynaecological or endobronchial applications, such testing only includes between one and three channels for normal treatment days. For days including implants utilizing most of the channels, measurements of the length for each channel individually become time consuming. A modification of a film-jig device, such as used for the radiograph in figure 4.3, allows testing of all channels as part of the daily check.

5.1.2.2. Source strength consistency. With an ionization chamber in a standard geometry (such as a well-type chamber or a Farmer chamber in a jig as discussed in the next section), compare a current reading with that measured at the time of calibration. For most well chambers, the calibration factor applies to the current scale, so the current reading comes about in the normal course of establishing source strength. For thimble chambers, the calibration factor usually relates to a

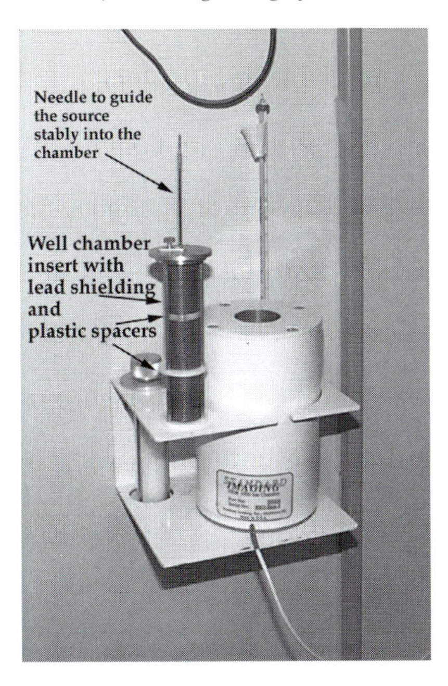

Figure 5.4. *Insert for a well chamber for assessment of source positioning.*

charge scale, and the current reading must be taken as an additional measurement specifically for the quality assurance. The current reading remains independent of the exposure control timer, reflecting a reading related to the source strength and the chamber response. If the length check uses the well chamber insert as described in section 5.1.2.1, the reading at point A serves also as the source strength verification measurement. Assuming the chamber receives periodic consistency testing using a long lived source that verifies consistent sensitivity (see section 2.1.1.3), any deviation from projected, decay-corrected readings (corrected for atmospheric density, that is, temperature and pressure) indicates a possible problem with the source strength. PASS: the current reading corrected for decay and atmospheric density falls within $\pm 1\%$ of that measured at the time of calibration; FAILURE: the corrected reading falls outside the acceptable limits.

5.1.2.3. Timer consistency. Correct dose delivery hinges on the proper operation of the timer controlling the exposure, of course. The proper operation does *not* depend on the timer accurately keeping normal clock time (that a minute indicated equals a true minute). The only important features of the timer are that it operates linearly (a 2 min setting runs twice as long as a 1 min setting) and that its operation remains consistent over time (either as real time or the timer's version). Initial evaluation of the timer during the calibration should cover a range of timer settings as encountered in treatments. This range generally runs from 0.1 s to about 200 s.

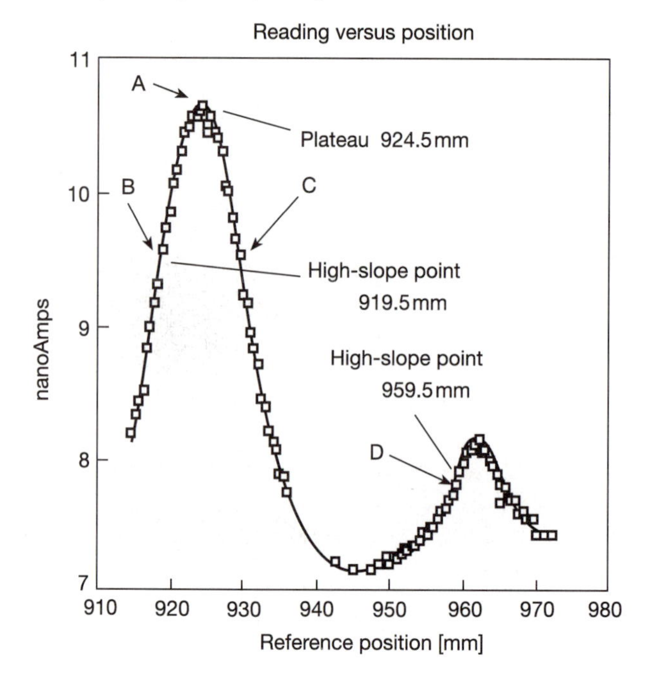

Figure 5.5. *Reading as the HDR source passes through the insert in the well chamber.*

The daily quality assurance generally need only check two times to evaluate timer consistency and linearity. The uncertainty in the measurement should be of the order of a few tenths of a second in order to check the shorter times.

One method for checking the timer compares time settings to intervals measured with a stopwatch. As an easy way to set up such a measurement, program the unit for several dwell positions in a row with the same dwell time, for example 3 s each. With a catheter attached, run the source out and listen through the intercom as the source advances. Each step makes a sharp sound, and with identical times, establishes a rhythm. Following the progress over several steps to adjust for the rhythm simplifies starting and stopping the stopwatch in the same part of the change in step. While for the longer times measurement with a stopwatch provides a good check for consistency, for the shorter times this method mostly provides information about the operator's response time. Again, discrepancy between the actual time on the stopwatch and that set on the timer does not indicate a problem, as long as the ratio of times remains constant.

A better technique for checking the timer observes the reading produced in a radiation detector as a function of the timer settings. To evaluate the timer, the measurement system must respond linearly to radiation dose and the setup provide a stable and reproducible geometry. Ionization chambers satisfy this requirement

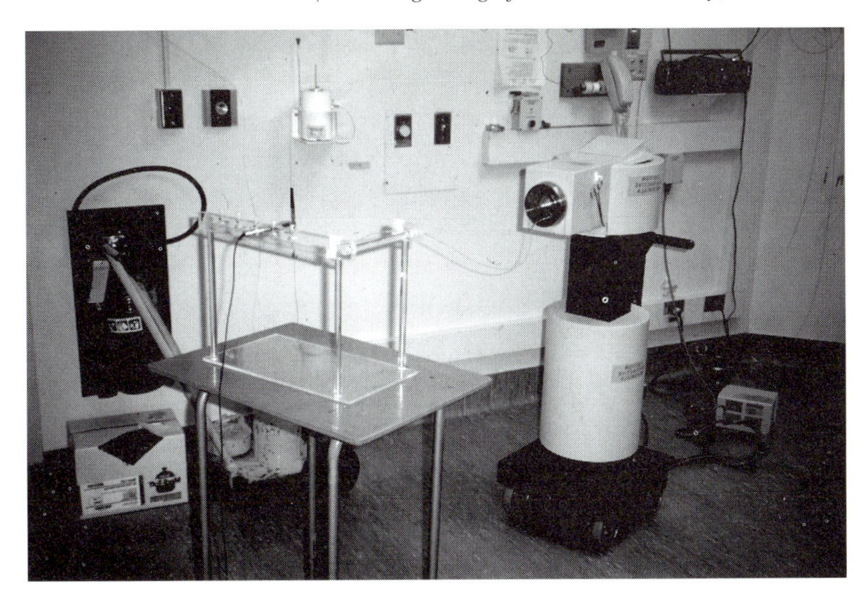

Figure 5.6. *Calibration setup for a high dose rate treatment unit using a thimble-type ionization chamber.*

if fixed in a reproducible geometry with respect to the source. They also provide a precision of the order of 0.1%—considerably higher than a person controlling a stopwatch for most meaningful dwell times. Diode-based dosimetry systems also can perform in this capacity as long as their response is checked for consistency approximately monthly on some other radiation source of an independently known calibration.

Something as simple as two holes in a block of Styrofoam™, one for a catheter and the other for an ionization chamber, serves well to fix the geometry for the timer evaluation. Figure 5.6 shows a more elaborate measurement jig. For daily use, a well-type ionization chamber provides the quickest and easiest arrangement. If a central shielding device as described above for checking the length is used in a well chamber, the reading at point A integrated over time works well to check just the timer since it remains fairly insensitive to minor changes in length positioning.

While the stopwatch approach measures just the timer's increments, the radiological measurements include the source decay and the effects of transit time. For the most part, source decay seldom deviates from the expected (although it has happened and does bear watching as discussed in the previous section). The usual method for the daily radiological check of the timer entails making the measurement immediately after calibrating a new source. This measurement serves as the standard for all subsequent daily checks performed with that source. The expected reading is then tabulated for each day through the next source change

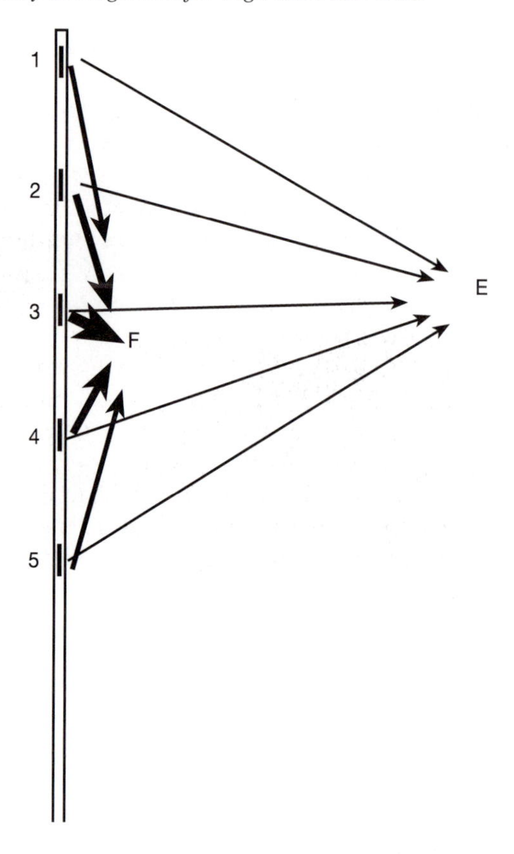

Figure 5.7. *The effect of distance from the source track on the relative dose received during transit. The widths of the arrows indicate the magnitude of the radiation the point receives while the source passes through the position at the arrow tail. For the transit dose at point F, position 3 contributes the dominant share, with less effect from positions 2 and 4, and relatively little from 1 and 5. For point E, all positions contribute approximately equally to the transit dose.*

by correcting the initial reading by the decay factor, $e^{-0.693t/73.8\ \mathrm{d}}$, where t equals the time in days since the initial reading. To further simplify daily evaluation of the readings, the tabulation of the expected reading can also include columns with this reading modified by the acceptable deviation. A common value for the acceptable daily deviation for the readings is $\pm 2\%$, although in practice deviations seldom exceed $\pm 1\%$ when using a well chamber. Figure 5.8 shows a sample chart with the acceptable upper and lower readings for a time period following an initial reading after the calibration. That the source follows the expected decay can be tested by taking readings in the same manner as for testing the timer except in the current mode, and comparing these to an 'expected reading' chart.

The effect of the transit time increases the reading beyond that due to the dwell time. With cobalt-60 units, whether the transit effect adds to or subtracts from the set exposure time depends on the setting of microswitches keyed to source movement that start the timer. No such switching or adjustments exist for HDR brachytherapy units, in part because the dose contribution to a point from the source in motion going to and returning from dwell positions depends on the geometry of the point with respect to the trajectory of the source. In general, the farther from the dwell position(s), the greater the fractional contribution to the dose due to transit effect. The reason for this can be seen in figure 5.7. Point E lies relatively far from the source track and the dwell position. As the source moves from the unit to the dwell, the distance to point E remains fairly similar, and the dose rate from each point along the way, being mostly determined by the inverse square law, remains constant. On the other hand, for point F, the source comes very close while at the dwell position, delivering a much higher dose rate than the relatively more distant points along most of the catheter. For a 10 s dwell time, most of the reading at point F comes from the source in the dwell position. However, with a source travel time of 4 s to come to the dwell position and another 4 s to return after the dwell time, 44% of the reading at E comes from the source in transit (8 s out of 18 s out of the housing, all with approximately the same dose rate). Thus, for checking timer consistency, points near the dwell position provide a more sensitive indication than those far away (although they amplify any minor variations in geometry). Nevertheless, transit time also is a parameter requiring evaluation for consistency. Variations in transit time not only affect patient dose, but portend serious machine failure. The readings from a Farmer-type ionization chamber positioned at about 15 cm from the dwell position in free space or a well-type ionization chamber combine sufficient signals from the source at the dwell for the set time and from source transit that a significant change in either yields a change in the reading, although without differentiating the cause of the failure. The role of the quality assurance procedure is to alert the operator to the existence of a problem: deviations call for further investigation.

Causes of deviations from the expected readings can come from:

(1) a change in the timer's period (so all times differ from what they used to be by a constant value);
(2) a change in the timer's linearity;
(3) a change in the transit time;
(4) the timer producing erratic durations;
(5) the transit time becoming erratic.

Differentiating among these possibilities requires successive elimination. Possibilities (4) and (5), erratic functions, show as wide variations for a series of measurements with the same geometry and dwell time. Unless the cause is gross, separating whether the timer or transit causes the erratic readings may not be possible through simple measurements. Erratic timer operation or transit may still fall within the uncertainty for evaluation using a stopwatch. Once identified, erratic functioning becomes a problem for the manufacturer's service technician. The timer linearity becomes the next feature to check. The daily check for linearity (discussed in the next section) assumes no change occurred for the transit time, and, thus, cannot be used in this case when the transit time comes into question. Instead, this situation requires a full check for linearity as performed at the time of calibration. This determination uses several measured readings, R_i, for various times t_i. Assuming that t_i increases as i increases, for adjacent pairs of readings,

$$\dot{R}_i = \frac{(R_{i+1} - R_i)}{t_{i+1} - t_i}. \tag{5.1}$$

By taking the difference between the readings and the times, the effect of source transit drops out of the equation. Making readings for five times from as short to as long as the electrometer scale allows yields four values for the reading rate. For a timer with adequate linearity, these values should differ by less than 0.5%. If the timer linearity proves satisfactory, the next step compares the value for the reading rate to that determined during calibration, correcting for source decay and atmospheric density. These should agree to within $\pm 1\%$. If the rate remains the same as at calibration, investigating the transit time comes next. Unfortunately, determining the actual transit time becomes difficult. The usual quantity of interest is the transit effect, the dose (or reading) resulting at a point due to the source while moving into the dwell position. This effect often finds expression as the equivalent time, t_ε, to deliver the dose (or cause the reading) were the source delivering that dose from the dwell position. During calibration, this time comes from

$$t_\varepsilon = \frac{R_i}{\dot{R}} - t_i \tag{5.2}$$

where the variables assume the same meaning as in equation (5.1). At this point, following the procedure outlined, t_i has been verified as free from significant error. However, if some doubt still exists for the accuracy of t_i an alternative method remains independent of t_i's accuracy. This method follows the multiple exposure technique for determining transit effect equivalent time, making two exposures: one, R_s, for a specified timer setting, and a second, R_m, for the same overall time, but broken into a number, m, of shorter exposures. The single exposure contains one transit, while the multiple contains m transits. The equivalent transit time becomes

$$t_\varepsilon = \frac{R_s - R_m}{\dot{R} - m\dot{R}}. \tag{5.3}$$

In this equation, the value for $\dot{R}$ comes from the decay-corrected value established at the time of calibration, again corrected for atmospheric density. The calculated equivalent transit time then can be compared to that determined during the calibration.

5.1.2.4. Timer linearity. If the initial test for timer consistency passed, a second reading evaluates linearity, making a reading using a set dwell time significantly different, usually by a factor of two, from the initial reading. Assuming a timer setting half of the timer consistency check for the linearity test, the reading will not decrease by a half. Both exposures include an equal contribution from the equivalent transit time, which keeps the shorter reading from dropping to half of the longer. The deviation from linearity, ζ, equals

$$\zeta = \frac{(R_{t1}/R_{t2})}{[(t1+t_{\varepsilon})/(t2+t_{\varepsilon})]} \tag{5.4}$$

where R_{t1} equals the reading for time $t1$, R_{t2} equals the reading for time $t2$, and t_{ε} equals the equivalent transit time. PASS: ζ falls between 0.99 and 1.01; FAILURE: ζ falls outside these limits.

5.1.2.5. Source strength value. Check that the value for the source strength in the treatment unit equals the initial value corrected for source decay. PASS: the treatment unit contains the correct value for the source strength $\pm 0.5\%$; FAILURE: the value in the treatment unit computer differs from that calculated by more than 0.5%.

5.1.2.6. Date and time. Check that the date and time in the treatment unit match the correct date and time. Some units compare the source strength in the treatment unit with that from the treatment planning system and adjust the dwell times for any differences. This correction also applies to any treatment programs stored and recalled. The unit determines the strength beginning from an entry at the time of calibration, which the computer then corrects for decay based on the days since calibration. Thus, an incorrect date can result in erroneous dwell times. Particularly likely and serious errors occur when the unit expects the date in either American or European format, and the operator enters the opposite. While this seldom causes a problem if entry of the date occurs only at calibration, problems arise for units requiring dates with treatment programs, or when the clock and calendar require updating because of daylight saving changes. PASS: the date is correct and the time matches the local time to within an hour; FAILURE: the date is incorrect or in the wrong format, or the time is different from the local time by more than an hour. Note that most units use the time only to gate when to change the source strength, not to calculate the strength based on the time of treatment. Thus, if the computer updates the source strength twice daily, say at noon and midnight, were the clock off by one hour and the computer updated the source strength early, the strength would only be different during the hour between when it did and when it would have changed. Assuming that the source strength values truly apply to the midpoint of the interval (6 p.m. and 6 a.m.), by the changing times, the strength contains an error of a quarter of a per cent. The hour error in the changing time means that instead of being 0.25% lower than the true strength, that in the unit exceeds the true strength by 0.3%. One hour later the unit's strength (which has not changed in that hour, nor will it for another 11 hours) exceeds the true strength by 0.25% as it normally would at that point were the time correctly set. Fairly large errors in the clock time still result in minimal errors in the source strength used for a treatment.

5.1.2.7. Transfer media. Check that the unit accepts correctly programs from the normal media used to transfer treatment information from the treatment-planning computer to the treatment unit. This may already have been accomplished by saving some of the programs used in previous checks on the media. For example, saving the dwell positions and times for the 'source positioning accuracy' checks saves time programming the units every morning. Recalling these from the medium serves to verify the operation of the medium. PASS: the treatment unit accepts the program correctly from each relevant medium; FAILURE: the unit refuses the program, or reads incorrect parameters from a medium. Note that if the treatment unit should adjust the dwell times for source decay, that also requires verification.

5.2. INITIAL CHECKS

The initial checks occur with each source change, but also contain some items that require periodic testing, but with frequencies more on the order of months rather than days. The evaluations performed with installation of a new source include:

(I) Safety checks:

(a) All of the safety checks performed with the daily checks.
(b) Backup batteries for room radiation monitors.
(c) Backup batteries for the unit.

(II) Dosimetry checks:

(a) Calibration of source strength.
(b) Determine equivalent transit time.
(c) Find the standard length.
(d) Enter the data into the computers.
(e) Record the initial data for daily checks.
(f) Create the limits table for the periodic checks.

(III) Ancillary device checks:

(a) Radiographic marker positions.
(b) Transfer tubes and adapters.

5.2.1. Safety checks

5.2.1.1. All of the safety checks performed with the periodic checks. This set of checks simply follows the list of safety checks performed at the beginning of each treatment day.

5.2.1.2. In-room radiation monitor backup batteries. These batteries provide power to the room monitor in case of a power failure. While all HDR treatment units contain their own backup batteries (see next entry) to retract the source in case of power failure, several situations could result in the loss of power to the room monitors with the source still extended. For example, a power surge could burn all the electronics in the treatment unit leaving the source extended, and simply knock out the power to the monitor (not that far-fetched). To check these batteries, unplug the detectors from the wall socket, and perform the normal *room monitor* check described in section 5.1.1. PASS: the monitor signals the presence of the source as with normal operation; FAILURE: the monitor fails to detect the source, or make audio or visual indications of the presence of radiation with the source extended.

5.2.1.3. Treatment unit backup batteries. The backup batteries for the unit provide power to retract the source in case of a power failure, and to hold the history of the treatment delivered up to the time of failure allowing resumption of treatment following restoration of power. To check these batteries, initiate a treatment programme lasting about one minute. *Part 1:* approximately halfway through the programme, pull the circuit breaker (or remove the fuse) for the line powering the unit. PASS: the source retracts before the unit shuts down; FAILURE: the unit shuts down with the source extended[1]. *Part 2:* restore the power to the unit and try to reinitiate the treatment programme. PASS: the unit picks up the programme where it left off at the time of the power failure, and completes the treatment; FAILURE: the unit either loses the programme or 'forgets' what was executed before the power loss.

5.2.2. *Dosimetry checks*

5.2.2.1. Calibration. As discussed in chapter 1, details on source calibration fall beyond the scope of this text. Briefly, the simplest and surest method for source strength calibration uses a well-type chamber itself calibrated at a standards laboratory (Goetsch *et al* 1992, DeWerd 1995). The calibration factor for the chamber usually gives the source strength per unit current reading. The well chamber could be the same one as used for low dose rate sources, but the scales used on the readout electrometer probably differ. The calibration factor for the same chamber will differ for low dose rate and high dose rate ^{192}Ir sources because of differences in source construction. The uncertainty in source strength calibrations using such chambers usually runs around 1% from national standards and 5% from absolute measures of energy absorbed per unit mass. These uncertainties provide

[1] Some units may calculate the power required to finish the treatment and retract the source, compare that to the power reserve in the batteries, and decide whether to complete the treatment before shutting down or simply retract the source. If the unit completes the treatment when challenged by the power loss, set a much longer dwell time and try the test again.

better values for patient treatments than manufacturers' values, which sometimes carry uncertainties of 10%.

If a positional accuracy insert is used in the well chamber to verify the proper length, obtaining the calibration factor also for point A (figure 5.5) saves time, and the possible mistake of performing the calibration with the insert accidentally in the well. Intentionally performing the calibration with the insert requires a little more care in positioning the source in the centre of the point A plateau, but eliminates switching inserts.

Measurements with a thimble chamber form an alternative to the use of a well chamber for source strength determination. Goetsch *et al* (1991) describe the procedure, which becomes long and involved, but certainly feasible at any facility. The method described makes readings at several distances from the source, and solves the simultaneous equations for the unknowns of offset in distance, room scatter and source strength.

5.2.2.2. Timer linearity. Several readings (often five) taken for various set times covering the range allowed by calibrated readout scales serve as the basis for evaluating the linearity of the timer. The use of the positional accuracy test tool extends the range of readings through the attenuation of the lead cylinders, after determining the ratio of the readings at point A and at a minimum in the response curve. As described with the periodic test of timer linearity, the readings in order of the time set compare the differences in readings for the difference in the time set, using equation (5.1). The reading rate, $\dot{R}$, should remain constant, and the maximum variation of $\dot{R}$ from the average of the four values gives an indication of the limits of the timer linearity. Values for $\dot{R}$ that change monotonically with increasing time set imply that longer times should be tested. If the readings have been taken in a well chamber, and the longest reading pushes the high end of the readout device, further experiments can be performed using a thimble chamber at a distance that allows measurements at longer times. Variations in $\dot{R}$ should remain less than 1% over all times tested.

5.2.2.3. Equivalent transit time. From the reading at any time setting and the value of $\dot{R}$, the equivalent transit time comes from equation (5.2). As mentioned above, this quantity varies with the measurement geometry, so measurements in the well chamber give different equivalent transit times from those made with a thimble chamber. Neither gives the true transit time, that is the time for the source to move from the shielded housing to the dwell position for the measurement, but any measured equivalent transit time serves as a baseline to watch for changes in the source transit.

5.2.2.4. 'Length' determination. This test follows the same procedure as with the periodic check.

5.2.2.5. Entry of data into computers. The new source strength must be entered into the treatment planning computer and the treatment unit computer. Some computer systems take a value entered in the afternoon and back-calculate the strength as of midnight. For calibrations performed in the morning but entered into the computer in the afternoon, this back-calculation gives an erroneous value by a half-day's decay. For such a situation, the measured source strength must be decayed by a half day before entry. Special attention needs to be paid to selecting the units for the strength, particularly if the system automatically defaults to given units, and also particularly during times of change in national calibration protocols.

5.2.2.6. Data for daily checks. With the measurements performed so far, the reading on the ionization chamber in the standard geometry or well chamber, $\dot{R}$, the equivalent transit time and the value for the length to correspond to the first dummy marker provide the baseline information for comparisons for the periodic checks, and should be recorded in a convenient place for reference.

5.2.2.7. Creation of periodic check limit table. Following all of the dosimetric evaluations, enter the values in the spreadsheet that provides the guidance limits for the periodic checks. (Figure 5.8 shows an example spreadsheet.)

5.2.3. Ancillary device checks

5.2.3.1. Radiographic marker positions. Correct positioning of the source in the target volume depends on not only the proper source length, but also the radiographic markers falling in the standard locations. Through use, the marker cables become bent, stretched and kinked, and the markers along the cable shift. With each calibration, these cables should be checked for proper placement of all the markers. A simple jig, shown in figure 5.9, facilitates these checks. The flange on one end holds the 'head' of the marker cable in place at the zero of the scale that runs along the long bar, where a scale allows direct readings of the position of each marker. Severely bent cables should be retired and replaced.

5.2.3.2. Transfer tubes and adapters. Some units base source positioning on an assumed length on the tubes (transfer tubes) that lead the source from the unit into the treatment appliance (applicators or needles). For units with this type of positioning, the length of the transfer tubes requires measurement to assure conformance with the assumed length. At this time, the physical condition of the tubes also should be noted, and the proper operation of any gates that prevent the source passing without a complete and closed applicator in place.

Date	Allowed reading limits		Source strength	Date	Allowed reading limits		Source strength
1/5/99	2.33	2.43	9.89	2/24/99	1.46	1.52	6.18
1/6/99	2.31	2.40	9.80	2/25/99	1.44	1.50	6.13
1/7/99	2.29	2.38	9.71	2/26/99	1.43	1.49	6.07
1/8/99	2.27	2.36	9.62	2/27/99	1.42	1.48	6.01
1/9/99	2.25	2.34	9.53	2/28/99	1.40	1.46	5.96
1/10/99	2.23	2.32	9.44	3/1/99	1.39	1.45	5.90
1/11/99	2.20	2.29	9.35	3/2/99	1.38	1.43	5.85
1/12/99	2.18	2.27	9.26	3/3/99	1.37	1.42	5.79
1/13/99	2.16	2.25	9.17	3/4/99	1.35	1.41	5.74
1/14/99	2.14	2.23	9.09	3/5/99	1.34	1.39	5.68
1/15/99	2.12	2.21	9.00	3/6/99	1.33	1.38	5.63
1/16/99	2.10	2.19	8.92	3/7/99	1.32	1.37	5.58
1/17/99	2.08	2.17	8.84	3/8/99	1.30	1.36	5.53
1/18/99	2.06	2.15	8.75	3/9/99	1.29	1.34	5.47
1/19/99	2.05	2.13	8.67	3/10/99	1.28	1.33	5.42
1/20/99	2.03	2.11	8.59	3/11/99	1.27	1.32	5.37
1/21/99	2.01	2.09	8.51	3/12/99	1.26	1.31	5.32
1/22/99	1.99	2.07	8.43	3/13/99	1.24	1.29	5.27
1/23/99	1.97	2.05	8.35	3/14/99	1.23	1.28	5.22
1/24/99	1.95	2.03	8.27	3/15/99	1.22	1.27	5.17
1/25/99	1.93	2.01	8.20	3/16/99	1.21	1.26	5.13
1/26/99	1.91	1.99	8.12	3/17/99	1.20	1.25	5.08
1/27/99	1.90	1.97	8.04	3/18/99	1.19	1.23	5.03
1/28/99	1.88	1.96	7.97	3/19/99	1.18	1.22	4.98
1/29/99	1.86	1.94	7.89	3/20/99	1.16	1.21	4.94
1/30/99	1.84	1.92	7.82	3/21/99	1.15	1.20	4.89
1/31/99	1.83	1.90	7.75	3/22/99	1.14	1.19	4.84
2/1/99	1.81	1.88	7.68	3/23/99	1.13	1.18	4.80
2/2/99	1.79	1.87	7.60	3/24/99	1.12	1.17	4.75
2/3/99	1.78	1.85	7.53	3/25/99	1.11	1.16	4.71
2/4/99	1.76	1.83	7.46	3/26/99	1.10	1.15	4.67
2/5/99	1.74	1.81	7.39	3/27/99	1.09	1.13	4.62
2/6/99	1.73	1.80	7.32	3/28/99	1.08	1.12	4.58
2/7/99	1.71	1.78	7.25	3/29/99	1.07	1.11	4.54
2/8/99	1.69	1.76	7.19	3/30/99	1.06	1.10	4.49
2/9/99	1.68	1.75	7.12	3/31/99	1.05	1.09	4.45
2/10/99	1.66	1.73	7.05	4/1/99	1.04	1.08	4.41
2/11/99	1.65	1.72	6.99	4/2/99	1.03	1.07	4.37
2/12/99	1.63	1.70	6.92	4/3/99	1.02	1.06	4.33
2/13/99	1.62	1.68	6.86	4/4/99	1.01	1.05	4.29
2/14/99	1.60	1.67	6.79	4/5/99	1.00	1.04	4.25
2/15/99	1.59	1.65	6.73	4/6/99	0.99	1.03	4.21
2/16/99	1.57	1.64	6.67	4/7/99	0.98	1.02	4.17
2/17/99	1.56	1.62	6.60	4/8/99	0.97	1.01	4.13
2/18/99	1.54	1.61	6.54	4/9/99	0.96	1.00	4.09
2/19/99	1.53	1.59	6.48	4/10/99	0.96	0.99	4.05
2/20/99	1.51	1.58	6.42	4/11/99	0.95	0.99	4.02
2/21/99	1.50	1.56	6.36	4/12/99	0.94	0.98	3.98
2/22/99	1.49	1.55	6.30	4/13/99	0.93	0.97	3.94
2/23/99	1.47	1.53	6.24	4/14/99	0.92	0.96	3.90

Figure 5.8. *Example of a data sheet made at the time of source calibration for guidance during the daily checks. The sheet contains the limits for the daily readings on the well chamber and the source strength for the day.*

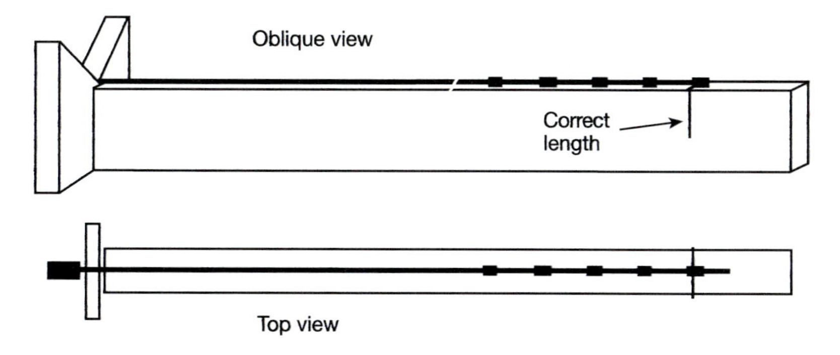

Figure 5.9. *A schematic of a jig to assist in checking that the radiographic markers fall at the correct length from a hub.*

5.3. ACCEPTANCE TESTING AND COMMISSIONING

In addition to performing all the procedures of the initial checks, when the HDR unit first arrives, some additional tests evaluate particular aspects of the unit, all dealing with safety features. Some of these tests require the use of a non-active cable in place of the true source cable.

(I) Operation.
(II) Detection of obstructions.
(III) Source retraction mechanisms.

(a) Emergency.
(b) Manual.

(IV) Loss of source.

5.3.1. Operation

Check that the source programmed for each channel correctly executes the programme, with particular attention to the location of each dwell position. The tests would be performed as described above. PASS: the source goes to the correct dwell positions at the correct distance along the catheter or needle; FAILURE: the source fails to pass into a catheter or go to the programmed dwell positions or positions the dwells at the wrong locations along the catheter.

5.3.2. Detection of obstructions

Cut the end from an endobronchial applicator and attach it to the treatment unit. Program the non-active 'source' cable to a position beyond the cut end. *Part 1:* as the source cable moves out of the end of the catheter, grab the cable (forceps may be necessary since the cable slides easily through fingers) together with the catheter so as to retard the cable movement. PASS: the unit detects the increased resistance, retracts the source cable and prints a notice of the problem; FAILURE: the unit continues to try to move the source cable forward. *Part 2:* assuming the success of Part 1, reprogram the unit and initiate the source run. Allow the source cable to move to the most distal position and again grab the cable. PASS: once the time comes for the source to retract, it detects the increased resistance and initiates an emergency retraction and prints a notice of the problem; FAILURE: the unit simply continues to try to retract the source in the normal fashion with no notice of a problem.

5.3.3. Source retraction—automatic (emergency)

This part of the test was just performed with part 2 of the previous test.

5.3.4. Source retraction—manual

Shut down all power to the unit and manually extend the source cable by moving the drive belt to the cable reel. (Alternatively, program the unit to send the source cable beyond the cut end of the catheter and initiate the program. Remove the connections to the backup batteries—unless that initiates an emergency retraction—and then cut all power to the unit.) With an assistant holding the cable by hand (not with forceps) manually crank the source into the shielded housing. PASS: the manual retraction mechanism successfully retracts the cable; FAILURE: the manual traction system fails to retract the source cable.

5.3.5. Loss of source

Program the unit to send the source cable beyond the cut end of the catheter and initiate the program. While the source dwells at the most distal position, cut the equivalent of the source capsule off the end of the cable. PASS: when the source cable retracts, the unit detects the difference in the cable length going out and coming back, signals an alarm, and prints a notice of the problem; FAILURE: the unit retracts the source without notice of a problem.

5.4. CHECKS DURING TREATMENT

During the treatment, the operator must remain alert, and aware of the condition of the source and the patient. Since the treatment takes only a few minutes, the operator should observe the following:

(1) the patient, watching for any movement that could affect the accuracy of the treatment or the patient's safety;
(2) the progress of the source so that at any time the operator knows what channel the source is treating, and that the source progresses at the appropriate times.

The operator must be ready at any moment to respond to emergency situations, or to halt the treatment if anything seems wrong or questionable.

5.5. CHECKS AFTER TREATMENT

At the end of a treatment, the first action performed by the operator verifies complete source retraction. This procedure entails several steps:

Verifying completed treatment per control console

The treatment monitor indicates termination of treatment and arrival of the source in the shielded container. The operator should be aware of any problem detected by the treatment unit before entering the room.

Observing the room monitors

Following source retraction, the warning lights by the treatment room door extinguish.

Measuring the radiation levels at the patient

Using the hand-held detector, check near the treatment site in the patient for elevated radiation levels indicative of the source, or part of the source, left *in situ*.

Measuring the radiation levels at the treatment unit

Using the hand-held detector check for the absence of high radiation levels around the unit. In some units, the source path enters a tube in the shielded housing before the path curves to obscure any straight-line rays. A source stalled at this point in the tube projects a pencil-like beam, usually in the forward direction. Reading with the detector in the location of this possible beam tests for a narrow, intense beam to which the room monitors might not respond.

5.6. SUMMARY

Figure 5.10 shows a form used at the University of Wisconsin for recording the information required for daily and initial tests. Because of regulations by the United States Nuclear Regulatory Commission, the unit also requires monthly calibration. Figure 5.11 shows the patient log, where each row records the treatments for a patient. This form assists in reminding the operator to verify the patient's identity and record the treatment in the patient's record book. Most of the tests discussed in this chapter, other than acceptance tests, fall on these forms.

	Initial Calibration	Daily QA		Week 1				
Date		Day		M	T	W	R	F
Source Number		Date						
RSR No.		Time						
Disposal No.		Initials						
Initials		Cable Attachment						
Time		Door Interlock/TV/Intercom						
Chamber		Warning Lights						
Electrometer		Hand GM Check						
Calibration Factor	U/10-7 A	Door Interrupt						
Atmospheric Factor		GM Sound/Red Light						
Measured Activity	U	Emergency Off						
Manufacturer's		Treatment Interrupt						
Activity	U	Timer Ends Tx						
Manufacturer's		Cable Length Check						
Measured		Source Retraction						
Linearity		Reading Pt B						
10 s reading	$\times 10^{-7}$ C	Reading Pt A						
D/s	$\times 10^{-7}$ C/s	Reading Pt C						
15 s reading	$\times 10^{-7}$ C	Reading Pt D						
D/s	$\times 10^{-7}$ C/s	Ratio B/A						
30 s reading	$\times 10^{-7}$ C	Ratio C/A						
D/s	$\times 10^{-7}$ C/s	Ratio D/A						
60 s reading	$\times 10^{-7}$ C	Length						
Variation in D	$\times 10^{-7}$ C/s	Temperature						
Transit	sec	Pressure						
Measured 3 s	sec	Ion Chamber Reading(10s)						
Dummy Length 1	mm	Corrected Reading						
Dummy Length 2	mm	Upper/Lower Limit						
Dummy Length 3	mm	Ion Chamber Reading(5s)						
GM Backup Battery		Source Retraction						
Dummies length should =1020mm ±2mm		Source	Pt 1					
		Retraction	2					
Calibration measurements recorded		Checks	3					
in the Week 1 Column			4					
			5					
Transit equation is			6					
t=10 s reading/(10* current)			7					
		Unit Activity						
		Comments						

Figure 5.10. *Sample check form to guide HDR unit testing.*

APPENDIX 5A. PULSED DOSE RATE UNITS

Pulsed dose rate (PDR) brachytherapy attempts to combine the best features of low and high dose rate brachytherapy. The main drawback of HDR brachytherapy involves the increased relative biological sensitivity of normal tissues compared to tumours at high delivery rates. Fractionation of radiation doses reduces that effect, as does the delivery at low dose rates. On the other hand, HDR, stepping source configurations allow optimization of the dose distribution to a much greater extent than with conventional LDR brachytherapy applications. Pulsed dose rate units use a single, stepping source to produce optimized dose distributions, but spread the dose over a duration similar to LDR treatments by delivering the dose in short pulses, repeated frequently over the course of treatment. For each pulse the source follows the program through the entire appliance, and then retracts until the next pulse. Commonly, each pulse begins hourly. While combining the desirable features of HDR and LDR approaches, PDR also suffers from the lack of immobilization characteristic of LDR applications, and the inability to hold

PT NAME	MR #	TYPE TX*		Tx 1	Tx 2	Tx 3	Tx 4	Tx 5
			Date					
			Pt ID**					
			Chart					
			Bill					
		Dictated	Time ck p Tx					
DISEASE	STAGE	Chart Copied	Pri Physicist					
PT NAME	MR #	TYPE TX*		Tx 1	Tx 2	Tx 3	Tx 4	Tx 5
			Date					
			Pt ID**					
			Chart					
			Bill					
		Dictated	Time ck p Tx					
DISEASE	STAGE	Chart Copied	Pri Physicist					
PT NAME	MR #	TYPE TX*		Tx 1	Tx 2	Tx 3	Tx 4	Tx 5
			Date					
			Pt ID**					
			Chart					
			Bill					
		Dictated	Time ck p Tx					
DISEASE	STAGE	Chart Copied	Pri Physicist					
PT NAME	MR #	TYPE TX*		Tx 1	Tx 2	Tx 3	Tx 4	Tx 5
			Date					
			Pt ID**					
			Chart					
			Bill					
		Dictated	Time ck p Tx					
DISEASE	STAGE	Chart Copied	Pri Physicist					
PT NAME	MR #	TYPE TX*		Tx 1	Tx 2	Tx 3	Tx 4	Tx 5
			Date					
			Pt ID**					
			Chart					
			Bill					
		Dictated	Time ck p Tx					
DISEASE	STAGE	Chart Copied	Pri Physicist					

* T&O:Tandem & Ovoids, T&C: Tandem & Cylinders, O: Ovoids, Endob: Endobronch, Endoes: Endoesoph, Temp: Template, Int: Interstitial

** Patient ID Verification: N-ask name; B-ask birthdate; A-ask address; I-ID bracelet or MR #; P-compare photo HDR-98-006.0

Figure 5.11. *HDR patient log.*

normal structures away from the appliance—an advantage of HDR treatments. For a more detailed discussion of the physics of PDR treatment, see Williamson (1995).

The PDR unit closely resembles an HDR unit, and functions similarly. While the source for the analogous HDR unit contains seven pellets of ^{192}Ir in a row, that for the PDR unit contains only one. The reduced activity of the source acts as protection for the patient should the source fail to retract following a pulse. Unlike an HDR treatment where the operator remains at the controls during the entire treatment, following initiation of a PDR treatment, the physician and physicist leave the area (and at night, go home), and the nursing staff go about their other duties, checking periodically on the patient. Thus, a source lodged in an applicator in a patient may dwell *in situ* for some time before detection and rectification. The diminished activity reduces the dose the patient receives before source removal in an emergency. While failure of source retraction rarely happens with HDR units, each treatment with PDR presents about 50 more opportunities for retraction problems. Because of the wear on the unit, preventive maintenance takes on a greatly enhanced importance.

For the most part, quality management for PDR units follows that for HDR units, except for two additional items.

Interpulse period (performed monthly)

Program the unit for a normal, but short, treatment, with a short interpulse period, and measure the interval with a stopwatch. Reprogram the unit with a longer interpulse duration and repeat the measurement. PASS: the measured interpulse duration matches that programmed; FAILURE: either or both measured times differ from that programmed. A failure leads to assessment of the linearity of the interpulse timer and of any offset.

Assessment of transit dose (performed per patient)

While really part of the treatment planning assessment, its applicability to PDR places this discussion here. Normally, the time taken moving sources into an applicator remains negligible compared to the total treatment time. Even for HDR treatments, the dose due to the source moving through the applicator only becomes important for implants with many catheters. However, in some situations the extra dose becomes substantial. Houdek presents an example for an HDR prostate implant where the transit effect increased the dose in the target by 7%, and to the rectum by 19%! (Hudak in Thomadsen *et al* 1994). The relative importance of the transit effect increases with distance from the source track (Bastin *et al* 1993). Because of the many fractions during treatment, PDR brachytherapy spends much more time with the source travelling through, and to, the appliance, delivering extra dose. Fortunately, the extra dose also depends on the absolute strength of the source, and the reduced source strength compared to an HDR source also protects the patient, to some extent, from transit dose. Once the source comes from the unit to the first dwell position, the time travelling between positions becomes subsumed in the dwell time at the arrival position. Most of the transit dose comes from the return of the source through the channel after completing all the dwells in a track.

Calculation of the transit dose requires knowledge of the velocity of the source passing through the appliance and transfer tubes. This can be determined by videotaping the source movement through a clear catheter against a ruler in the background, and noting the change in position on a frame by frame playback. While the speed may reach 50 cm s^{-1}, there is an acceleration phase beginning as the source leaves the last-treated dwell position. For the calculation, each centimetre along a catheter may be assigned a transit dwell time equal to the inverse of the velocity at that distance from the beginning of the return. For example, if the velocity at a given location approximates 20 cm s^{-1}, assigning all the time it takes the source to pass through the centimetre centred on that location gives it a dwell time of

$$\frac{1\ \text{cm}}{20\ \text{cm s}^{-1}} = 0.05\ \text{s}.$$

Using this approximation for the dwell times for dwell positions separated by 1 cm steps allows estimation of the transit dose to critical structures using the same algorithm as the normal calculations. The calculation includes dwells not just in the appliance, but also in the transfer tubes. However, as the source passes into the tubes, the velocity increases, reducing the equivalent transit dwell time, and the distances begin to reduce the doses to negligible levels. For the most part, even with the multiple fractions, the transit dose remains quite low for most realistic situations, although, if all the transfer tubes for a large implant together pass over the patient's leg, for example, the dose might become high enough to consider spreading the tubes or redirecting the orientation of the tubes.

CHAPTER 6

QUALITY MANAGEMENT FOR HIGH DOSE RATE TREATMENT PLANS

More than any other form of brachytherapy, high dose rate (HDR) brachytherapy relies on extensive quality assurance to prevent serious errors in patients' treatment plans from becoming serious executed errors. Continuing the discussion begun in chapter 5 in more detail, several features of HDR brachytherapy combine to make this procedure particularly hazardous:

- *Few fractions.* Most HDR treatment courses deliver the therapy in two to five fractions. Unlike normal external-beam treatments, an error in a single fraction constitutes a significant fraction of the total dose.
- *Large fractions.* While external-beam treatments typically deliver about 2 Gy per fraction, because of the low number of fractions, HDR brachytherapy treatments often use fractions of 6 Gy or more. Some treatment protocols use five fractions of 9 Gy for early-stage cervical cancer (Stitt *et al* 1992) and two fractions of 16 Gy for prophylactic vaginal cuff irradiation following hysterectomy for endometrial carcinoma (Noyes *et al* 1995a). With these larger fraction sizes, slight errors in doses greatly increase the probability for complication.
- *Rapid pace.* Particularly with gynaecological applications, from the time of insertion through localization and planning to the actual treatment, all the activities proceed very quickly to minimize the time the patient lies waiting. The urgency arises for several reasons:
  - *Potential thromboses.* Many of the patients, particularly those with compromised blood flow and the heavy set, present a nontrivial probability for phlebitis and thromboses arising in the legs while lying in the dorsal lithotimy position, with legs elevated in stirrups. The modern practice of using pneumatic socks that periodically inflate to keep the blood moving reduces, but does not eliminate the hazard.

- *Patient discomfort.* Patients lying on their back in stirrups for long periods also experience back pain and general discomfort, again particularly the heavy set. This discomfort often leads to the third reason . . .
- *Patient movement.* As patients feel uncomfortable, they begin to move. Movement between the localization procedure and delivery of treatment compromises the accuracy of the dosimetry in the treatment plan. As discussed below, accuracy becomes much more important for HDR brachytherapy than with LDR.

- *Short treatment times.* The treatment proper usually takes less than a half-hour, and most commonly only about ten minutes. While LDR treatments allow a review of the plan and loading after insertion of the sources, with HDR the delivery is likely to be complete before an error becomes apparent if all reviews are not finished before commencement of treatment.
- *Large numbers of data to process.* Compared to LDR brachytherapy treatment planning, HDR plans often require more input data. The increased data not only arise from the larger number of HDR dwell positions typically used compared to the number of LDR sources that would be used for the same treatment, but mostly due to the fundamentally different approaches treatment planning takes for the two modalities. Usually LDR brachytherapy treatment planning proceeds by selecting strengths for the sources in established positions. If the dose distribution fails to satisfy some criteria, the planner tries different strengths, although the number of possibilities remains quite low. Even with ^{192}Ir sources in ribbons or ^{125}I sources in a permanent implant, optimizing may take many iterations; the input data remain limited with few changes in loading between iteration. HDR brachytherapy treatment planning reverses the process, with the planning specifying the desired dose at various locations and the computer determining the times for each dwell position. This process may still require iteration as the planner varies some of the optimization parameters or manually changes dwell times to create the optimum plan. The HDR brachytherapy approach requires a better understanding up front than that of LDR brachytherapy of the demands of the treatment (i.e., the doses to various points in the patient, desired shape of the isodose surfaces). A complete HDR brachytherapy treatment plan requires input of more than 350 informational items (not keystrokes, but full quantities), offering many possible entries for errors to creep into the planning.

The quality assurance programme should employ both approaches to error reduction: error prevention and error interception.

6.1. ERROR PREVENTION

For HDR brachytherapy treatment planning, error prevention amounts to reduction in the probability of errors in the input data entered into the treatment planning computer or treatment unit. Applying the three techniques presented in chapter 1 helps prevent errors from entering the data.

6.1.1. Protocols

A protocol simply provides a formalized, standard set of expectations and procedures. When all persons involved follow a protocol, deviations from the protocol stand out as possible mistakes and items calling for investigation before proceeding to execution of the treatment. Because planning and execution of HDR treatments proceed so quickly, formal protocols become extremely important since much of the information passes between persons with assumptions about meanings and intents. Take for example dose protocols. For HDR treatments, the protocols need not only doses, but fractionation schemas. If the doses vary depending on certain treatment factors, the planner must understand the formalism to determine the variations in dose. As mentioned in chapter 1, doses prescribed for a given cancer often depend on the stage, grade or location of the disease, but the dose for a given fraction can vary based on the conditions for that application. The person performing the dosimetry must be able to distinguish between a mistaken prescription and one varying due to conditions—or at least know when to ask. Another protocol may dictate at which doses to critical structures the planner notifies the physician and asks whether the prescription should change.

6.1.2. Forms

As with other quality control tools, forms become more important for HDR treatments than for LDR brachytherapy. Not only do the forms remind the person checking a plan of items to review, but help speed the evaluation procedure, and let the reviewers know when they are finished.

6.1.3. Independent second person

An independent second person finds several roles in HDR treatments. As noted in chapter 1, a second person watching the identification of catheters can prevent mistakes during that process. During data entry into the planning computer a second person watching the operator enter data can call attention to errors as they occur for instant correction. The University of Wisconsin has made it a policy to have two certified operators[1] attending all HDR planning sessions, and found that the 'observer' often notices errors and prompts corrections.

6.2. ERROR DETECTION

These quality control checks compare the results of the treatment plan with objective measures that distinguish the correctness of the final product. Again, as with

[1] Certified implies trained for and tested in the performance of a task. See chapter 1.

error detection, independence of the evaluations forms an important factor in the value of the review.

The model used at the University of Wisconsin serves as an illustration. This model uses three separate reviews of HDR treatment plans, except as noted below.

6.2.1. Independent physicist's review

A different physicist performs the review from the one who served as observer for the dosimetrist performing the treatment plan. Sometimes during treatment planning, a difficult or confusing issue arises, and the planners discuss methods of handling the problem. The participants in the discussion can (and not uncommonly do) convince themselves that the solution arrived at is correct and carries no resultant detriments to the patient's treatment. A third person, not privy to the discussion, seeing the result may question the appropriateness of the planners' actions and the plan (often rightly), while those involved with the planning would not. Examples of common situations of this type include interpretation of the intended target for dose specification and desired active dwell positions used.

Many smaller departments run with a single physicist or cannot free a second physicist for a review of the plan (although such reviews only take about 10 min). In this case, the 'independence' must come from imposed systematic and objective checks. The physicist performing the review should acknowledge the compromise in 'objectivity' of judgements regarding the plan, and requestion any unusual results or procedures used during planning.

Task Group 59 of the American Association of Physicists in Medicine (1998) suggest that the physicist performing the evaluation have the following materials available for checking:

(A) General:

(1) Physician prescription form, completed, signed and dated.
(2) HDR physics form, filled out, signed and dated.
(3) All simulator radiographs.
(4) Patient chart, including all previous HDR plans.
(5) Isodose plot of current plan, signed by physician.
(6) Printout of current plan.
(7) Printout from the transfer medium program for the current plan.

(B) Radiographs:

(1) Proper set(s) of radiographs.
(2) Correct radiograph orientation on the digitizer.
(3) Correct numbering of catheters and dwell positions.
(4) Information transferred from initial planner to radiographs.
(5) All desired patient dose points properly defined and marked.
(6) Correct dose origin in the same place on all radiographs.

(C) Plan and plot:

(1) Patient identification.
(2) Source activities from the table and from the planning system agree to within 0.5%.
(3) Source strength, ID, date of calibration.
(4) Correct number of catheters.
(5) Step size.
(6) Source geometry reconstruction.
(7) Start and end dwell positions.
(8) Correct indexer length for each catheter.
(9) Radiograph orientation, magnifications and film-to-source distance.
(10) Applicator point coordinates.
(11) Correct dose optimization points on printout and plot.
(12) Correct prescription-point dose on printout and plot.
(13) Reasonable plot for the positions of the patient points.
(14) Check on the printout that the doses to the patient points are reasonable.
(15) Reasonable agreement with previous plan.
(16) Smooth dummy wire travel.

The heart of the review lies in checks likely to find aberrant treatment plans. This function warrants very detailed consideration, found below in section 6.3.

Figure 6.1 shows a form used for review of a gynaecological cancer treatment using a tandem and ovoids. The checks fall into several general categories.

6.2.1.1. Dose prescription. This section looks at whether the treatment plan accurately satisfies the dose prescription. High dose rate brachytherapy prescriptions should include the target site, the total number of fractions, the dose per fraction, the total dose for the therapy and any external-beam treatment doses that will be part of the therapy. To evaluate that the plan correctly satisfies the prescription requires review of the following:

(1) *Comparison of the delivered dose to the prescribed dose*

Doses for HDR plans may be normalized in one of several ways. One method pegs the dose at a specified location, either with respect to patient anatomy or to the applicator. Treatment planning based on imaging studies (e.g., CT or ultrasound) often utilizes such dose specification. Superposition of the dose distribution on the images, either as isodose curves on serial slices or as three-dimensional surfaces together with rendered shells of the patient's organs or the target volume, allows evaluation of adequate coverage. With radiographs as the localization medium, patient points need high contrast for positive positioning. Surgical clips serve well for defining the location of points of interest. Contrast material in body cavities seldom gives very precise indications of anatomic points. Part of the increase in

HDR DOSIMETRY CHECK

Cervical Cancer Treatments Using Tandem and Ovoids

Date __________________ MR# ________________________

Patient ____ Fraction No. ________ of _______

Stage:____ Dose/Fraction from protocol:__________________

1. Location and Dose Checks

____ a. Dose for this fraction on Rx __________ Gy Average dose to applicator points __________ Gy

____ b Difference between right and left M and prescribed dose is less than 5%

____ c Distance of point M Cervical Stopper:__________ Middwell ovoids:_____________

Distance cephalad as defined in Rx __ ______ mm Distance cephalad on films ________ mm

Distance lateral as defined in Rx ________ mm Distance lateral on printout _______ mm

Distance lateral on coronal plane _______ mm

____ d Ovoid cap sizes

Rt Visible marker ______ Rt size ________ mm Rt Distance to vaginal dose points ________ mm

Lt Visible marker ______ Lt size ________ mm Lt Distance to vaginal dose points ________ mm

____ e Dose percentile to vaginal surface ___________ % of Rx dose = __________ Gy and isodose lines on the plan fall on the vaginal surface

____ f Starting dwell for tandem on plan corresponds to start indicated on film Dwell :________________

____ g Bladder ____________ Gy (___________%) Rectum _____________ Gy (____________%)

__________ Physician alerted if > 70%

2. Time Checks:

____ a Time index for dwell 1 cm from first dwell Index ________ Posted range ______ to ______

____ b Time index for total time Index ________ Posted range ______ to ______

____ c Total Time Index from previous treatment Index ________ Agree within 5%

3. Card/Treatment Unit Check

____ a Rt. ovoid programmed to channel 1 Length for this channel ________

morning QA length _________

Lt. ovoid programmed to channel 2 Length for this channel ________

Tandem programmed to channel 3 Length for this channel ________

____ b Step size ________ 2.5mm ________ 5.0mm

____ c Patient's file has been saved.

4. Programming of the HDR Unit

____ Dwell times, positions, and length on print out match that from the computer planning

All the appropriate checks above prove satisfactory

________________________ ________________________ ________________________

Physicist Time Date

Figure 6.1. *(a) A check to assist a physicist in the review of an HDR treatment using tandem and ovoids for cervical cancer.*

PHYSICIAN'S HDR DOSIMETRY CHECKLIST

Cervical Cancer Treatments - Tandem and Ovoids

Check in the box indicates parameter is correct.

Patient ________________________ Medical Record No.____________

Date ________________________ Fraction No. ________of__________

_____ Prescribed dose per fraction of _ Gy matches that from protocol for stage _, given as _ Gy

_____ Prescribed dose is shown as red line on plan.

_____ Prescribed dose falls on points M on coronal plane.__ Points M fall 20 mm from the origin.

_____ Vaginal optimization points each fall _____ mm from the center of the vaginal sources, corresponding to the radius of the ovoid caps of _____ mm.

_____ Dose to the vaginal surface (_____% x _____ Gy @ Pt. M) of _____ Gy falls on the vaginal optimization points, and is shown as a line on the dose distributions.

_____ Maximum doses to the bladder, _____ Gy, and rectum, _____ Gy have been noted.

_____ The treatment time of _______ sec (_______ minutes) corresponds to the patient's previous treatments to within $\pm$5% when corrected for source decay and differences in applications. (Previous time _____ sec x decay correction _____ = projected time ______ sec)

_____ "Length" on the program, that is, the distance to dwell position No. 1, corresponds to the posted length, unless there is an offset

__ Check here if there is an offset = __ mm Length = posted length - offset = ___ mm

_____ Starting dwell for tandem on plan corresponds to start indicated on film.

_____ Dwell times on the treatment machine program match those from the planning computer

_____ On visual check, the applicator appears not to have moved since placement.

All appropriate checks above prove satisfactory.

________________________ __________ __________

Resident Time Date

________________________ __________ __________

RSC Authorized Attending Physician Time Date

DECAY CORRECTION TABLE

Days	Correction	Days	Correction	Days	Correction
1	1.009	8	1.078	15	1.151
2	1.019	9	1.088	16	1.162
3	1.028	10	1.098	17	1.173
4	1.038	11	1.109	18	1.184
5	1.048	12	1.119	19	1.195
6	1.058	13	1.129	20	1.206
7	1.068	14	1.140	21	1.217

Figure 6.1. *(b) A check to assist a physician in the review of an HDR treatment using tandem and ovoids for cervical cancer.*

uncertainty comes from the difficulty in identifying the same point on two radiographic views, particularly for irregularly shaped objects. Something as simple as contrast in a Foley bulb in the bladder that almost forms a sphere presents a case where the posterior aspect in anterioposterior and lateral views allows fairly accurate reconstruction (see figure 6.2). However, even for this example, locating the closest point to some particular part of an applicator becomes much more difficult. Skeletal anatomy sometimes presents unique aspects facilitating identification in separate radiographic views. Unfortunately, dose to skeletal points seldom forms the basis for the treatment prescription, although some target anatomies, such as many lymph chains, follow the skeleton in many parts of the body. However, individual patients present with wide variations in the actual location of lymph chains, as seen in patients with lymph angiograms.

Most intracavitary insertions use simple radiographs for localization, on which most target anatomy remains invisible. Assuming that the application conforms well to the anatomy of interest (such as a tandem and ovoids in a patient's uterus and vagina), specifying the dose to a position relative to the applicator serves the same function, except with an additional uncertainty. The Manchester point A used for dose specification forms an example of such an applicator-based treatment. Potash *et al* (1995) discusses the uncertainty, noting that, depending on the definition used (he notes about 19 in the literature), point A often falls in a region of high dose gradient near the ovoids, where small differences in the actual placement of the point can make very large changes in the amount of radiation the patient receives. Some of the definitions move the point more cephalad where the isodose surfaces tend to follow the tandem, and variations in the actual position of the point make little difference to the treatment. Many temporary implants localized through radiographs only display the needles or sources, with the dose specification based on the implanted volume. These implants also are applicator-based treatments. Regardless of the nature of the applicator-based points, specification of the dose to one or an average of several serves as a surrogate for specification to the actual patient anatomy, but assumes a highly conformal (and skilful) implant.

A special subset of applicator-based dose specifications arises with optimized applications. The optimization process often uses many points as locations for specific values of dose. The computer then adjusts the dwell times for the source to satisfy as best as possible the requested doses at the optimization points. Assuming that the optimization points represent some significance with respect to the patient, the absolute dose can be specified to the average of the optimization points.

Some optimization routines use no specification points, and nonoptimized treatment may use predetermined dwell times or relative dwell weights. Such applications become much like LDR implants, and the selection of dose follows the same guidance.

For an intracavitary insertion a review of the dose entails checking not only the dose at the particular prescription point or points, but that the dose to the remainder of the target receives the appropriate coverage. Often the optimization

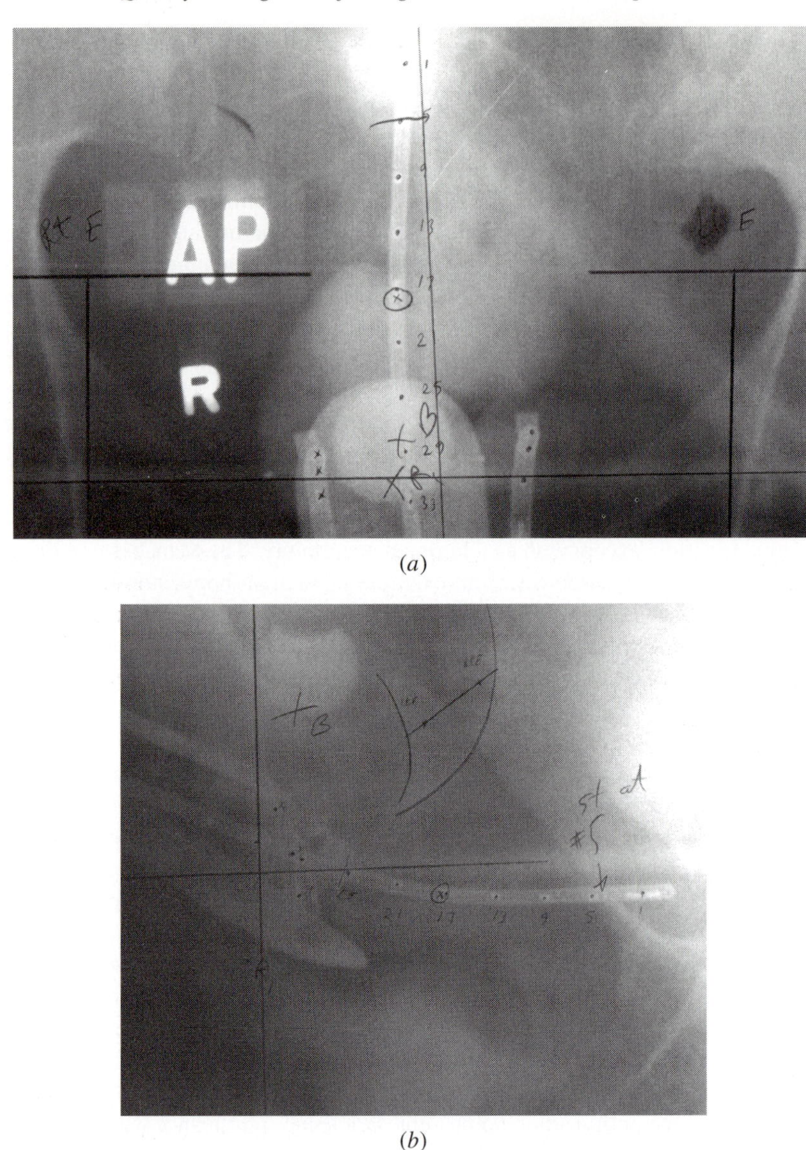

(*a*)

(*b*)

Figure 6.2. *An example of locating structures in two orthogonal radiographs.*

points serve this role, having been placed around the applicator to establish the treatment isodose surface (or surfaces if points of various dose values were used). A check of the resultant doses to these points measures the conformality of the treatment. For nonoptimized cases, or for cases with too few optimization points to provide an adequate review (a situation to be avoided, since so few points provide insufficient information for the optimization engine), the evaluation requires comparing the position of the treatment isodose curve on relevant planes to the location of the target. The person performing this review needs to understand the three-dimensional shape of the target. Cervical cancer treatments often only describe the desired dose distribution as 'pear-shaped', with no more quantitative details. While optimization *requires* quantitative descriptions of the target volumes, *any* planning should decide *a priori* the target in sufficient detail to evaluate whether the objectives are met by the plan.

(2) *Evaluation of the uniformity of the dose*

For the intracavitary treatment under review with the form in figure 6.1, the evaluation for treatment uniformity amounts simply to comparing the dose to the applicator point on the right and left side. For the tandem and ovoid, the tandem might deviate toward one side, or one ovoid might fall closer to the tandem than the other. A rigid appliance avoids this potential problem. Interstitial implants present a greater challenge in evaluating the 'uniformity'. The chapter on low dose rate treatment planning discusses means for evaluating implant uniformity, and all of that material applies to high dose rate implants as well.

The approach to assessment of the uniformity of an intraluminal application (e.g., endobronchial and endoesophageal insertions) falls somewhat between the interstitial and the intracavitary evaluations. The goal of the treatment often targets delivery of the prescribed dose to a constant distance from the source catheter. While many endoesophageal insertions closely approximate a straight line, appropriate optimization often achieves this goal to within a few per cent over the target length in all directions. Bronchi, on the other hand, usually curve through the treatment length. The curvature produces higher doses on the inside of the curve and lower doses on the outside. Since most bronchi only curve so to remain in one plane in the regions most often treated, the doses on the sides normal to the curve should be approximately equal. A test for uniformity in such cases first looks at the doses on the two sides normal to the curvature. In the central cut perpendicular to the treatment catheter, the doses on the two sides at the treatment radius should not differ from their average by more than 3%. For an optimized application, the dose for points along the length of the catheter in these normal directions should fall within 4% of the average of all the doses, excluding points opposite dwell positions within half the treatment radius of the ends of the treatment length. (See figure 6.3 for an illustration of these points.) The doses in the plane of curvature will probably deviate from the average by much more than 4%, often by 15%. This difference is unavoidable; optimization cannot improve on, or reduce the variation. The reason for checking the extent of asymmetry is to alert the physician, who must decide to continue with the treatment or not. The doses along the catheter in the plane of curvature likewise will vary considerably, even on a single side.

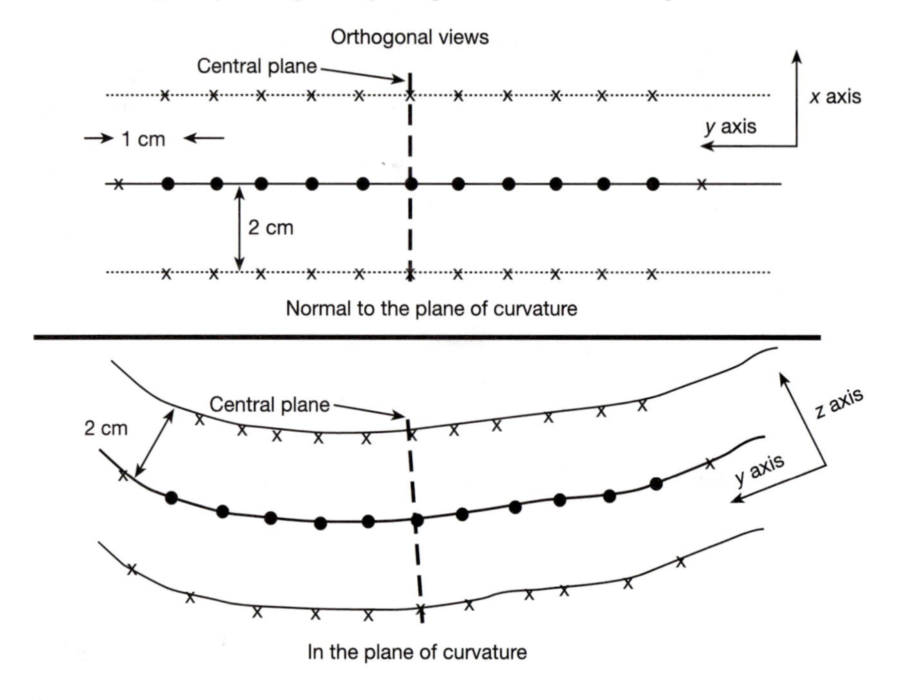

Figure 6.3. *Illustration of optimization points (×) for an endobronchial application. Dwell positions are indicated by the filled circles.*

(3) *Location of the prescribed dose*

HDR treatment planning systems often calculate the relative dwell weights based on the optimization routine, and then set the absolute dwell times from the dose at a specified location. Erroneous times result from either an incorrect specified dose *or* a dose specified at the wrong location. The location may be with respect to the applicator, the patient or the localization image system, but in any case the specification refers to a point in three dimensions. Part of specifying the point frequently includes establishing the coordinate system. Again, errors in determining the coordinate system translate directly into errors in the point location. While tolerances in the dose evaluation merely reflect the acceptable allowed limits on the dose, acceptable limits on the location of the prescription point depend on the gradient of the dose distribution near the prescription point.

(4) *Dose distribution/differentials in dose*

Treatments often require various doses through different parts of the application. For example, an intracavitary insertion may specify the dose to point A and along the tandem as 100% and that to the vaginal surface as 138%. An interstitial implant may deliver 6 Gy per fraction to a large volume with 8 Gy per fraction to a smaller internal volume. The differentials in dose should be set before creating the treatment plan; however, just because the dose distribution fails to achieve the desired differentials does not automatically indicate an unsatisfactory plan. Some prescribed distributions prove physically impossible, either due to uniformity requests or differentials that defy physical laws. That notwithstanding, the prescribed distribution serves as a measure for evaluation of the generated plan, and in some cases the failure to achieve the prescribed distribution *does* indicate that the procedure would not benefit the patient and should be aborted. Two important radiobiological considerations affect the decisions relating to the dose distribution:

(a) Changes in the relative dose distribution with HDR brachytherapy produce larger biological differences than with LDR brachytherapy. The curve relating the biological effectiveness and dose (figure 6.4) presents a larger slope with HDR brachytherapy than LDR brachytherapy. Thus, the variations in dose allowable with LDR brachytherapy can produce unacceptable complications with HDR brachytherapy. The second consideration comes from the same biological phenomenon.

(b) HDR brachytherapy presents a much smaller margin for errors in dose than LDR brachytherapy. Particularly for treatment doses very near normal tissue tolerance, small errors can push the tissues past tolerance. (See discussion below.)

Each type of treatment, and often individual patients' needs, dictate the allowed deviations from the desired differentials in the dose distribution. Still, an effective assessment requires setting the desired differentials *a priori* and evaluation of the adequacy *a posteriori*.

Part of verification of the dose distribution may include review of the optimization technique used. While the final measure of adequacy for the dose distribution rests on whether the prescribed dose falls in the desired locations, in reality, most desired distributions, explicitly or implicitly, describe a function impossible to achieve. In such cases, the question becomes how closely the planned distribution matches the desired, and could the planned distribution provide a *better* match with a different optimization approach or parameters? For example, with the Nucletron planning system's polynomial optimization routine, the factor called the dwell weight gradient factor governs the allowed variability of the dwell times: large values of the dwell weight gradient factor produce more uniform dwell times. Low values of this factor allow more flexibility in dwell times and can produce better matches between the planned and desired dose distributions. Very low values may generate some negative dwell times. Thus, the best plan requires a balance for the dwell weight gradient factor. A different algorithm in the Nucletron system, geometric optimization, may underweight the end dwell positions of catheters, in which case the polynomial optimization may achieve better the goals of the prescription. The reviewer should evaluate the quality of the match between the desired and generated distribution, with consideration of how the computer arrived at the plan.

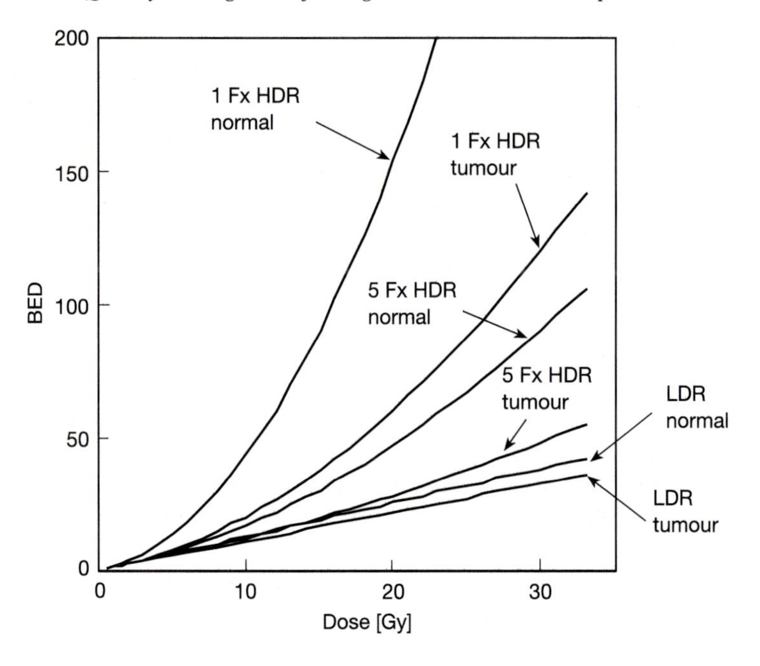

Figure 6.4. *The biologically effective dose (BED) as a function of dose for several treatment modalities. For the high dose rate modalities, the number before 'Fx' indicates the number of fractions over which the dose is delivered.*

Verification also entails checking the identity of the appliance in the patient. As discussed in chapter 3 and shown in figure 3.1, important parts of many appliances fail to be visualized on radiographs. Special markers assist in this identification.

(5) *Dose distribution begins and ends correctly along the catheter*

The previous check evaluated the dose distribution proper, in isolation from the mechanism to place that distribution in the patient. This review looks into the positioning of the distribution. One of the most common errors in high dose rate brachytherapy is applying the correct dose distribution to the wrong position along the catheters or needles. One problem with a general discussion of this aspect of the treatment arises because each of the treatment units approaches this positioning differently. Even a given unit may have different methods of positioning the dwell positions along the catheter. Figure 6.3 shows an example of a small endobronchial treatment, comprised of 20 consecutive dwell positions with a 0.25 cm step between, beginning 3 cm from the first possible dwell position as indicated by the first x-ray marker along the localization train. The Nucletron microSelectron allows execution of this plan in two ways. For the first, the program identifies the length for the treatment (i.e., the distance from the face of the unit to the first active dwell position) as 3 cm shorter than that length to have the source coincide with the first x-ray marker. The distance to the x-ray marker forms one of the daily checks during the morning QA of the unit (see chapter 5). In this example, the length equals 965 mm. The second method maintains the coincidence between the first dwell position and the first x-ray marker. In this case, the first dwell position activated falls (3 cm/0.25 cm/step) = 12 positions *after* the first dwell position, or position 13. Either method requires verification that the positioning was programmed correctly.

The distribution could be in error if selection of the dwell positions activated based on radiographs of the x-ray markers assumed a given step size, yet the size specified during planning differed. Such an error becomes apparent by checking the length of the treatment on a printout of the plan compared to that indicated by reference to the x-ray markers (or simply a specified length).

An error in the magnification of the input images translates into errors in the absolute dose delivered to the prescription point or volume, but such an error remains subtle and difficult to find. However, that error usually appears more apparent as an incorrect treatment length. A check of the treatment length may uncover any of several mistakes.

Most planning systems will not allow the input of radiographic coordinates with a mismatch between the type of reconstruction algorithm and the images used (for example, attempting to use stereo-shift films after declaring that orthogonal films would be used). The computer program should note large differences between the reconstructed coordinates from the two sets of images, and prevent progression through the program. However, some combinations of images *could* fall within reasonable differences when interpreted as belonging to a different reconstruction nature, particularly with a single catheter. To detect such an error, the reviewer *should* check compatibility between the relationship of the images and the reconstruction algorithm. Unfortunately, treatment-planning systems seldom print this information. Usually, such mismatches produce concurrent errors in the length (and possibly width) of the dose distribution.

6.2.1.2. Dose to normal structures. Normal structures often limit the dose any treatment can safely deliver. Reviews of any plan evaluate the dose to nearby normal structures. The biological considerations may accentuate or diminish the effect of the dose. The biological effectiveness of a high dose rate dose, D, delivered in N fractions of d grey per fraction, probably follows the linear–quadratic equation,

$$\mathrm{BED}_{HDR} = D\left(1 + \frac{d}{(\alpha/\beta)}\right). \tag{6.1}$$

For normal structures the (α/β) ratio[1] probably approximates 3, or 2 for

[1] The (α/β) describes the ratio of the exponential coefficients in the survival equation $S = \exp(\alpha D + \beta D^2)$.

some nerve tissues. The ratio for tumours varies considerably, but calculations often use a value of 10. The biologically effective dose to normal tissues relative to the tumour treatment dose, the fractional biological dose, FBD, defined as

$$\mathrm{FBD} = \frac{\mathrm{BED}_n}{\mathrm{BED}_t} \tag{6.2}$$

becomes

$$\mathrm{FBD}_{HDR} = \frac{D_n}{D_t}\left[\frac{1 + d_n/(\alpha/\beta)_n}{1 + d_t/(\alpha/\beta)_t}\right] \tag{6.3}$$

where the subscripts n and t represent the quantities evaluated for normal tissue and tumour, respectively.

Evaluation of the effect to normal structures most often requires calculating the FBD and comparing that to the value acceptable for an LDR treatment, since most practitioners at this time have a better feel for the more conventional therapy. The LDR equation for BED for most temporary implants lasting more than 15 h, delivered at a dose rate $\dot{D}$, approximates to

$$\mathrm{BED}_{LDR} = D\left[1 + \frac{\dot{D}}{\mu(\alpha/\beta)}(1 - 1/\mu t)\right] \tag{6.4}$$

ignoring cellular repopulation, where μ is the repair constant for sublethal cellular damage (taken as 0.462 h^{-1}) and t is the treatment duration. Substituting equation (6.4) into equation (6.2) with the appropriate values for tumour and normal tissues gives the expected FBD for a comparable LDR treatment, to which the FBD for the HDR treatment must compare. Complicating this analysis, the ratio of the physical dose to normal tissue to that to tumour may not be the same for the HDR and LDR treatments, since the advantages of the high dose rate approach include the ability to hold normal structures away from the source in some instances where such would be infeasible with an LDR approach. Upon calculating the FBD for the HDR plan and comparing it to that expected were the treatment delivered at a low dose rate, the physician must decide whether delivering the treatment remains in the patient's best interest. Stitt *et al* (1992) calculate that as long as the physical dose to normal structures at risk in treatment of the uterine cervix remains less than 83% of the dose to the target (point M in their system), the probability of an increase in complication over LDR treatment becomes unlikely.

6.2.1.3. Programming All the information to deliver the plan just reviewed must be transferred to the treatment unit. Depending on the configuration of the planning and treatment units, this may be accomplished either through transfer using an electronic medium (e.g., network transfer, disk or cartridge or programming cards), or by programming the treatment unit manually. Manual programming obviously opens avenues for errors, and most users recognize the hazards. However, electronic transfer offers opportunities for analogous errors, albeit fewer, while setting up the transfer. Each manufacturer's unit presents different critical factors. In general the items to check include the following:

(1) *Program identification*

The program medium or file should carry the patient's name or similar identification, such as a medical record number, unless the program corresponds to a standard treatment, in which case the label should carry the name of the standard. Standards saved in the treatment unit must carry some identification in the file and printed when recalled to distinguish the program.

(2) *Date and data file information*

The various manufacturers' systems require the passage of different information between the planning computer and the treatment unit. Some information required by various systems includes:

- the date of plan generation (in the expected format, i.e., European or American),
- the source activity for the generated plan and
- the data file for the plan calculation.

The source activity in the planning system should be checked against a table generated at the time of source exchange and calibration, using an independent computer program.

(3) *Treatment length*

The exact method to designate where the treatment begins along the catheter depends on the treatment unit. Often, the specification refers to the distance to the first dwell position, regardless of whether it is used. The unit then keeps track of the active dwell positions relative to the designated dwell position number 1. Treatment planning systems often set the value for this parameter as the default standard length if not otherwise specified. This default feature accounts for a large fraction of accidents where the wrong part of the patient receives the dose.

Finding the correct length value becomes a crucial part of the treatment planning. For units that specify the length to the dwell position number 1, different methods apply depending on whether the catheter or needle track falls short of the standard default length. For treatments where the localization markers slide fully into the catheter so the flange of the marker train abuts the collar of the catheter adapter, the length to the first marker comes from procedures described in the previous chapter. To use this position as the dwell position number 1 simply entails entering the established distance to this marker as the length. To move this dose distribution so it begins elsewhere along the catheter requires determining the *offset* between the first marker and the marker corresponding to the new dwell position number 1. Subtracting this offset from the standard length gives the length for entry into the program.

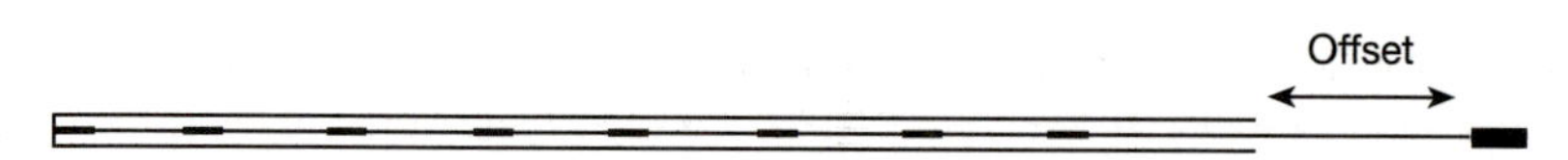

Figure 6.5. *Illustration of a measurement of the distance that should be subtracted (offset) from the standard length so the first dwell position falls at the end of the needle.*

Localization accuracy depends on simulating the actual treatment situation as closely as possibly. Several potential pitfalls lie in wait during this procedure. The person performing the measurement must ensure:

- *The use of the complete adapter.* Failure to place the adapter on an endobronchial catheter shifts the position of the markers by 16 mm. This relatively small shift easily goes unnoticed. For needles or gynaecological applicators, omitting the adapters or transfer tubes often leads to the marker train abutting the end of the needle or applicator, a shift in the position of the dose distribution of up to 1 cm for the gynaecological applicators and possibly more for interstitial needles.
- *The use of the correct combination of the transfer tubes, catheter lengths and marker train.* Assembling the applicators and transfer tubes completely up to where they insert into the treatment unit and using the corresponding marker trains displays on the localization radiographs the first dwell position using the standard length. This simple, and seeming obvious, procedure uncovers several potential problems. For example, accidentally using transfer tubes designed for treatment lengths of 1.5 m while assuming the length should be 1 m, using the 1 m long marker train would result in localization radiographs with no markers showing. (The markers would fall half a metre down the transfer tubes from the intended treatment site.) Some manufacturers market short marker trains for use with gynaecological applicators. These markers slide into an applicator only, at a point before the applicator connects to the transfer tubes. Using such markers not only fails to show mismatches between the transfer tubes and the intended treatment length, but also ignores any effects due to curvature of the transfer tubes, or mistaken identities of tubes.

Most interstitial implants use a different method for establishing dwell position 1. In these cases, because the length to the end of the track often falls considerably short of the standard distance, when inserted the marker train abuts the end of the needle or catheter (see figure 6.5). In this case, if dwell position 1 should fall at the end of the track, the distance comes from subtracting the excess length of the marker train (labelled as 'offset' in figure 6.5) from the standard length. Again, attaching the transfer tubes and using the standard marker train (as opposed to using a short train) assures that the length measured matches that for the application. A shift in the position of the dose distribution from the first dwell position is accomplished as before, by subtracting the desired offset from the measured distance to the first marker. Some manufacturers market a device that reads the length directly, eliminating the calculation as a potential source of error.

Evens *et al* (1993) noted that sending the source to the absolute bottom of a catheter, using the length indicated either by a ruler, marker offset, or the maximum length before the unit detects a blockage, results in errors in source positioning up to several millimetres for units using a set length for ejection. This occurs because the source cable flexes upon contact with the catheter end, and the flexing takes up some cable length. Thus, the unit registers that the source lies further than it truly does, and all subsequent dwell positions also include a shift.

All of the transfer tubes for use with interstitial needles *should* be of equal length, as should all the needles. However, the quality control in manufacture, particularly with needles, often results in variations of several millimetres. The needles require individual measurement at the time of treatment, or sorting by length into batches at receipt (and a system to keep the needles in the identified batches, and periodic checks). All interstitial cases using catheters require individual length measurements for all channels since the length of catheters varies considerably, and the catheters frequently are cut to convenient lengths after insertion into the patient.

Some manufacturers use an 'end-find' mode that sends the check cable down a channel until it reaches the end, and registers this distance. With this technique, QA on the length becomes part of the QA for the machine, rather than the plan.

Curvature affects the location of the source dwell positions compared to the marker train locations. This is discussed further in chapter 9, and figure 9.8 shows a comparison of some of the actual stops the source makes in the ring applicator compared to the marker locations. The differences arise because the markers all ride on a cable, and the mid-cable positions find support from both ends. The source, on the other hand, always falls at the end of the cable, with support on one end only. No commercial dosimetry program corrects for this effect, nor is consideration of this a usual part of the quality control. While the differences in the dose distribution remain small, in some cases this may be something to consider.

(4) *Step size*

If the unit allows different step sizes, the designation in the program must correspond to that in the plan.

(5) *Correct assignment*

The source track numbers in the planning system may not always correspond to the desired channel number. For example, the planning system may only allow sequential entry from the radiographs, yet during reconstruction and planning, advantages may become apparent in connecting and treating the tracks in a different order than entered. For example, for a two plane implant, entry into the planning system may be simplest from left to right on an AP projection, while connection might best proceed with all needles in one plane first followed by the needles in the second plane. Planning systems may or may not allow such a change in assignment from planning to programming. A common situation occurs when treating a tandem in a uterus with no other applicator components. For one manufacturer's system, the planning system may assign the tandem a track number of 1 because it is the first, and only, track in the plan, yet the treatment unit may interlock the tandem to only connect with channel 3 (channels 1 and 2 being assigned to the patient's right and left ovoid, respectively). Regardless of the intent, the reviewer must check that the program correctly assigns the source tracks to the proper channel. This check entails reviewing the dwell times for a track, and checking that they match those in the channel program.

6.2.2. Physician's review

A physician should also evaluate the treatment plan generated by the computer system. In this case, the question of 'independence' proves a little troublesome. On the one hand, just as with the physicist, an independent physician remains more likely to question some aspect of the application than a physician involved with the treatment. Such a person provides an excellent measure of quality review. However, an evaluation of the application requires considerable knowledge of the patient. By the time the physician performing the application relays all this information to the reviewing physician, it is not clear that the latter remains 'independent and unbiased'. Providing staffing for an independent physician review also becomes a problem for any but the largest facilities. Thus, this discussion assumes that the same physician performs the physician's review as performed the procedure, and that the review evaluates the correspondence between the generated plan and the desire of the treatment. In facilities where an independent physician performs the review, the review should include the appropriateness of the application.

The physician's check evaluates many of the same parameters as the physicist's check, but from a different perspective. The checks also provide redundancy, which, as noted before, does not imply useless repetition in quality assurance. The physician's checks include that:

(1) the prescribed dose matches the dose in the protocol;
(2) the correct dose occurs at the prescribed point or points, and that the prescribed points fall at the correct location;
(3) the dose distribution covers the volume as intended;
(4) the dose to critical organs remains within an acceptable range, given the patient's situation;
(5) the total treatment time corresponds to that from a previous treatment fraction, if applicable; and

(6) the length to the first dwell position matches that indicated on the localization images.

6.2.3. *Predelivery checks*

If all the above evaluations indicate that the generated treatment plan contains no significant errors and adequately satisfies the prescription, the treatment progresses to delivery. After programming the treatment unit and just prior to delivery, sets of predelivery reviews provide final verification of treatment parameters.

Patient identification

Patient verification proceeds as discussed in chapter 1.

Program identification

Programs for the treatment unit transferred via electronic media should be labelled in some way associated with the patient, such as by name or identification number. Even though later checks should uncover an erroneous program, at this point the label for the program should be verified. This becomes particularly important for 'standard plans', those plans used repeatedly for many patients, all receiving identical treatment. In these cases, should there be several standard plans available, calling the wrong plan becomes an easy mistake.

Program verification

The program loaded into the treatment unit must match that generated by the treatment planning system. This becomes particularly important for manually programmed treatments, but remains the last line of defence against erroneous treatments even for electronically programmed sessions. Important information to verify includes:

(1) the source strength compared to that on the treatment plan,
(2) the step size,
(3) the length to the first dwell position for each channel,
(4) each dwell position used and its respective dwell time and
(5) the total treatment time (as a redundancy check).

The date in the treatment unit should have been checked during the morning quality assurance procedure. The physician who inserted the applicator should make a last visual check that no gross movement occurred during the planning, review and programming time.

Under ideal situations (i.e., where staffing permits) the operator, the verification physicist and the physician all verify the program on the treatment unit. As a minimum, at least two persons knowledgeable of the unit operation must review the program. However, the operator maintains the final responsibility for assuring execution of the correct program on the correct patient.

Williamson (1991) describes a method for verification using autoradiographs. With this technique, a set of treatment needles, catheters or applicators identical to those in the patient are placed on a piece of film. The operator places the treatment unit into a special mode that steps through all of the active dwell positions, stopping for the same predetermined time for each (usually around 1 s when using Kodak^{TM} XV RediPak^{TM} film). Making an x-ray exposure of the setup with the marker trains in the applicator before or after running through the program creates the image of the applicators and dwell positions as a background for evaluation of the locations of the treatment dwell positions along the applicator. This check requires that the patient not be in the treatment room. Placement and localization for most interstitial and intraluminal applications takes place in other venues, leaving opportunity for performance of this test. Gynaecological applications sometimes have insertion and localization performed elsewhere, also permitting the autoradiography prior to treatment. However, gynaecological treatments frequently take neighbouring normal structures to doses very near tolerance, and successful therapy makes use of the ability to temporarily push these structures away from the applicator to a greater extent than tolerated under the longer duration of LDR treatments. Maintaining the apparatus that moves structures aside may be poorly tolerated by the patient during the additional time for the autoradiograph, or if not locked in place, the spacing apparatus may shift between localization radiographs and treatment execution due to motion from moving the patient. The motion also compromises the precision of the dose calculation, and the added uncertainty could translate into a higher fraction of patients with unexpected complications or recurrences. Thus, while the autoradiographs serve a valuable function in verifying that the program will execute using reasonable dwell positions located correctly along a source track, the assurance rarely justifies intentionally removing a patient before a gynaecological treatment.

Channel verification

All the preceding verifications prove futile if the catheters are connected to the wrong channel on the treatment unit. While different 'key' configurations on both ends of gynaecological-applicator transfer tubes often allow connections only between the correct part of the applicator and channel, none of the HDR units key channels for needles or catheters. Clear marking of the catheter or needles at the time of localization is essential (see chapter 9). To prevent accidental mixing between channels and needles, each channel should be connected completely to the unit one at a time, and verified by a second person. In addition to simply numbering the needles or catheters, a diagram made during localization or at the time of implantation not only assists during connection, but saves relocalization if the markings on the needles 'fall off'. Should questions arise as to the identity of catheters or needles, connection should not be based on guesswork. Relocalization, or at least verification fluoroscopy, should be employed, if appropriate.

6.3. INDICATORS OF 'REASONABLENESS'

The preceding checks concentrated on the questions of conformality between the plan and the prescription, and correct programming of the treatment unit to execute the plan. A big question remains: does the plan contain a serious blunder? The approaches in the literature to evaluate that question all fall into the realm of comparing facets of the treatment to some standard table or formula, with limits set around normal or typical plans. The assumption becomes that if the plan falls within the normal limits, the dose it delivers remains within acceptable tolerances. The various types of treatment all have different approaches to this test.

For any of the tests described below, calculations performed to verify the computer calculations *must not* use the output times and coordinates from the planning system simply entered into a second calculation program, or even a hand calculation. Such a test only checks that the calculation algorithms for the two systems agree—something that should have been checked during commissioning of the treatment planning system. Aside from detecting corruption of the program or data files (the former unlikely and the latter should be under control), such a procedure supports no barrier against the more common errors in digitization, specification or normalization. Re-entering the entire case into a different planning system (or, for that matter, the same planning system) provides a measure of the accuracy of the digitization and dose specification for the case, although, if the same person performs both plannings, erroneous decisions likely will be made twice. A second person performing the planning from scratch does supply item by item comparisons for all the important parameters of the plan. However, such duplication exacts a high cost in personnel time, and often the time also of a very uncomfortable patient. To speed the verification and reduce the staff expense, many groups developed consistency measures, as discussed below.

6.3.1. Gynaecological intracavitary insertions

Practitioners of low dose rate intracavitary treatments, even when prescribing the dose to points such as the Manchester point A, used mg h (or mg Ra eq h) as a check on treatment plans. Most treatments fell within a narrow range of mg h. Based on this, a first test of an HDR plan would look at the integrated reference air kerma, IRAK in U s (or, equivalently, total disintegrations as Bq s, or in conventional units, Ci s) of a plan. This simple approach requires two refinements. For LDR applications, the mg h varied with the stage of disease (or basically, the dose delivered). Since the total dwell times scale linearly with absolute dose, dividing the U s by the dose removes the dependence on the absolute dose prescribed. The mg h for LDR brachytherapy remained fairly constant in part because applications varied little. Most patients' uteri accommodated three of the relatively large LDR sources in the tandem, and one each in the ovoids. Occasionally a tandem in a long uterus held four, or a very short uterus two. Due to the small range of variation in the number of sources, the loading patterns remained constant also, with little ability to vary the application based on the needs of the patient. HDR applications vary widely in the number of dwell positions in the tandem. The number of dwells in the ovoids or ring remain the same, although the number of dwells used with a tandem in the uterus and cylinders around the tandem in the vagina vary depending on the length of the vagina to be treated. Williamson *et al* (1994) defined an 'index' for intracavitary insertions as

$$\text{total time index} = \frac{\text{sum of the dwell time} \times \text{strength of the source}}{\text{prescription dose} \times \text{number of dwell positions}}. \qquad (6.5)$$

This index gives the average IRAK over all dwell positions per unit dose.

The total time index relates to the integral dose to the patient, and approximately to prescribed dose for normal applications, but presents no information on the shape of the dose distribution. The doses near the ovoids depends in a complicated way on the dwells in the ovoids and tandem, yet doses near the tip of the tandem reflect little of the dwells in the ovoid. Examining a single dwell time near the tip of the tandem should give an indication that the distribution of the dwell times falls within a normal range. The very tip dwell time varies quite erratically, yet the dwell 1 cm inferior to the tip tends to be relatively stable. An index analogous to the total time index characterizes this dwell position,

$$\text{tip time index} = \frac{\text{dwell time 1 cm from tip} \times \text{strength of the source}}{\text{prescription dose}}. \qquad (6.6)$$

Table 6.1 presents values used for these indices at the University of Wisconsin. The limits for the total time index depend on the applicator used and the anatomy under treatment. The tip time index, falling a relatively long distance from the vagina where the applicators vary, depends only on the target.

After a patient's first fraction, subsequent treatments often assume very similar geometries. A second check for a major error in the calculations compares the values for the indices for a later fraction to those for the previous fraction. Variations up to 5% are considered acceptable. Deviations greater than 5% require investigation and a reasonable explanation before proceeding with the treatment. For a given patient, the tip time index can vary considerably between fractions when the plans use very low values for the dwell weight gradient factor (discussed above). The total time index remains fairly immune from effects of this factor since the factor mostly determines the distribution of dwell times rather than their sum. Comparison of the tip time index becomes unreliable for these low values of the dwell weight gradient factor, and this check is omitted. Other causes of changes in the indices between fractions include differences in the spread of the ovoids, changes in the length of the tandem in the uterus or the number of dwell positions used in the tandem or a change in the relationship between the tandem and the

Table 6.1. *Limiting values for the check indices in equations (6.5), (6.6) and (6.12). The values listed correspond roughly to two standard deviations for patients treated at the University of Wisconsin—Madison. Cases falling outside the limits need further evaluation.*

Applicator		Diameter of caps or cylinders (cm)	Limits in (Gy m^2 s)/(h Gy)	
			Lower limit	Upper limit
Tip time index	Tandem—any cervix ca case		0.139	0.180
Total time index	Tandem and ovoids	2.0	0.098	0.123
		2.5	0.123	0.147
	Tandems and cylinders	2.0	0.098	0.139
		2.5	0.114	0.147
		3.0	0.110	0.168
	Ovoids alone	2.0	0.086	0.098
		2.5	0.127	0.143
		3.0	0.184	0.196
Total time index′	Single catheter		0.270	0.315

ovoids. Obviously, a fixed applicator system obviates these changes, although it loses conformality with the patient's disease and anatomy. A check during later fractions presents the opportunity for comparison with a number of previous fractions, agreement with any of which provides some assurance that the current plan contains no blunders. Sufficiently large changes in the application on later fractions, such as changing from medium ovoids to small due to radiation-induced atrophy, leads to consideration of the plan in the manner of a first application, without precedent in that patient.

In this model, the physician's check for dwell time consistency only applies a comparison of the total dwell time to previous fractions. This comparison requires adjustment of the previous time for the decay of the source, using the table at the bottom of the form in figure 6.1(b). Thus corrected, the total times should agree to within 5%. The physician may need to make a slight correction for applications that differ by the number of dwell positions activated in the tandem. If the initial comparison falls outside the limits, the first step is to note the difference in the number of dwell positions. For example, the first application may have used a longer tandem and contain two more dwell positions than the second. The extra dwell positions are identified as being in the middle of the tandem. While in reality, none of the dwell positions actually constitute the extra positions, the ovoids dominate the dwell positions near the inferior end of the tandem, and the

Table 6.2. *Manual dose verification methods for optimized single-catheter applications using equation (6.7). (Reproduced from Ezzell and Luthmann 1995, with permission.)*

Expected values of dose index 10 cm from application in its central transverse plane	
Implant type	Expected range of dose index
Pulmonary	1.05–1.20 (lower if highly elongated, higher if curved)
Vaginal cylinder	1.10–1.20 (i.e. short single catheter)
Long oesophagus	0.95–1.10 (i.e. long single catheter)

tip end, as noted above, remains fairly constant. A longer tandem effectively just adds middle dwells. Thus to make the two applications more similar, two of the middle dwell times are subtracted from the total time of the first application before correcting for decay and comparing the time to the second application's. A similar subtraction of 'excess' dwell times corrects for a longer tandem in the second application. As with comparisons of total time indices, significant changes between two fractions call for treating the latter fraction as if it were a first fraction of a different treatment.

6.3.2. Intraluminal insertions

The most common intraluminal insertions include (in decreasing order of frequency) endobronchial, endoesophageal or transbiliary, with other applications comprising a small fraction. Most of these treatments use a single catheter, although two-catheter applications in the lung are not uncommon, and form the approach of choice in some facilities, sandwiching tumours when possible. While three-catheter endobronchial insertions arise, obtaining good dose distributions in these cases presents a considerable challenge to the best optimization routine (and operator), and restricts the patient's available airway. The discussion that follows considers consistency checks for single-catheter insertions. For multiple-catheter verification, see section 6.4.

Several different consistency checks for single-catheter applications appear in the literature. For optimized insertions, Ezzell and Luthmann (1995) calculate the dose to points 10 cm distant from the approximate centre of the sources, along a line approximately perpendicular to the catheter, in opposite directions. At that distance, the dose varies little with application geometry. They calculate a dose index, defined as

$$\text{dose index} = \frac{100(\text{dose}_{+10\,\text{cm}} + \text{dose}_{-10\,\text{cm}})/2}{\text{source strength} \times \text{total time}} \tag{6.7}$$

and compare it to the expected values in table 6.2.

This index proves sensitive for errors involving mismatches between intended dose and that specified in the plan, or between the assumed source strength and that in the computer, unless the person performing the review uses the values from the plan in the check. The check can also reveal algorithm changes (not a common problem in most facilities) or changes from normal in the distance from the catheter used for the target dose (a much more common problem). This approach fails to detect errors in the details of the dose distribution or in the dwell positions used, or mistakes in the digitization. Understanding the limitations of this index allows its use as a consistency check for those aspects of the treatment for which it applies.

An approach taken by Kubo (1992) and Kubo and Chin (1992) calculates the approximate total treatment time for applications prescribed at 10 mm from the centre of the catheter based on a linear fit, using the equation

$$\text{time [s]} = \frac{0.01[\text{s Ci/cGy mm}] \times \text{dose [Gy]} \times (2.67\,\text{length [mm]} + 78.6\,\text{mm})}{\text{source strength [Ci]}} \tag{6.8}$$

where 'length' equals the active length along the catheter. The authors suggest that this predicted time and that calculated by the planning system should agree to within 2%. For most single-catheter treatments, the prescription distance seldom approaches values equal to the active length. Thus, the dose tends to vary as the distance to the first power, in the absence of significant curvature. With that assumption, the equation above applied to other treatment distances becomes

$$\text{time [s]} = \frac{0.1[\text{s Ci/cGy}] \times \text{dose [Gy]} \times (2.67\,\text{length [mm]} + 78.6\,\text{mm})}{\text{source strength [Ci]} \times \text{prescription distance [mm]}}. \tag{6.9}$$

The group that developed equation (6.9) further modified the techniques to calculate the time for arbitrary lengths and treatment distances through interpolation between the times for 50 mm and 200 mm treatment lengths using a quadratic fit (Rogus *et al* 1988). With this model, the treatment time becomes

$$\begin{aligned}\text{time [s]} = {} & \left(\frac{\text{dose [Gy]} \times 2\text{Ci/Gy}}{\text{source strength [Ci]}}\right)\Bigg[(-5.24 + 8.80d + 0.263d^2) \\ & + \left(\frac{L - 50\text{ mm}}{L - 200\text{ mm}}\right)(-9.66 + 22.4d - 0.027d^2)\Bigg]\end{aligned} \tag{6.10}$$

where L = active length in mm and d = prescription distance, also in mm.

Ezzell (1994) developed a similar equation to predict treatment time based on film dosimetry. His equation for a dose prescribed to 10 mm follows

$$\text{time [s]} = \frac{\text{dose [Gy]}}{\text{source strength [Ci]}}\,\frac{L}{0.0537L + 23.09} \tag{6.11}$$

where again L equals the active length in mm. The relative dwell weights for use with this equation follow the pattern from one end of the catheter to the other of:

1.5, 1.4, 1.3, 1.2, 1.1, 1.0, 1.0, ..., 1.0, 1.0, 1.1, 1.2, 1.3, 1.4, 1.5.

The University of Wisconsin model approached evaluation of the total time for a single-catheter insertion in a manner similar to that for a gynaecological application, using a total time index defined as

$$\text{total time index}' = \frac{\text{sum of the dwell times [s]} \times \text{strength of the source [Ci]}}{\text{prescription dose [Gy]} \times \text{treatment length [cm]}} \frac{2\text{ cm}}{d} \quad (6.12)$$

where, in this case, the treatment length equals the distance from where the prescribed dose falls along the catheter on one end to where that dose intersects the catheter again on the other end. The variable d equals the treatment distance in centimetres. Table 6.1 gives values also for this index.

6.3.3. Interstitial implants

High dose rate implants differ physically very little from low dose rate versions used under the same situations. As a result, the same rules that guided the LDR implants apply to the HDR. Most implants approximate those described by some system, if for no other reason than some system applies to most reasonable approaches. One effective method to check an HDR implant, then, is to calculate the time predicted using the appropriate tables or graphs from a system that matches. Optimized implants form the norm since the planning system handles that operation with ease. The resulting differential loading simulates the approach of the Manchester system, which likewise attempted optimized uniformity through differential loading. Because such HDR implants parallel those following Manchester, the dosimetry tables from the latter provide effective predictions of the treatment time...with some modifications. The first modification reflects differences in the target volume. In the Manchester system, following the loading rules produced doses in the target volume within $\pm 10\%$ of the nominal (i.e. prescribed) dose. For a volume implant, the minimum dose tended to fall on the periphery, so the peripheral dose would be 10% lower than the prescribed dose. Generally, not only have most physicians moved to prescribing the dose at the periphery, but follow the lead of the Paris system and prescribe to isodose surfaces outside the confines of the implanted volume. For most volume implants, the Paris system's basal dose approximates the prescribed dose location in the Manchester system. While the Paris system defined the reference dose as 85% of the basal dose for a uniformly loaded implant, for an optimized implant using 90% maintains the prescribed dose at the same location. Thus, for a Manchester-type implant prescribed in the Paris-type fashion, the IRAK (equivalent to mg h times a constant) becomes that calculated using the Manchester table divided by 0.90. However, for HDR implants, the biological effectiveness accentuates differences in dose, so to achieve the same ratio of biological effect that the 85% produced with uniformly loaded LDR implants, optimized HDR implants should use a ratio of 0.93 instead. (See Thomadsen 1995b for a more detailed discussion.) The other, more obvious changes required to use the Manchester tables for HDR implants simply consist of modifying the units to match the current units for source strength and changing the

Table 6.3. *Quality assurance index values for regular volume implants, derived from the original Manchester tables, corrected by replacing 1000 R with 9 Gy. Conversion to Ci s by using the factor* 1.79×10^{-3} *Ci mg Ra eq*$^{-1}$, *conversion to air kerma rate by using* 7.225×10^{-3} *Gy* m^2 h^{-1} *mg Ra eq*$^{-1}$.

R_V for HDR Volume Implants					
Corrected and specificed to the periphery					
Volume	R_V		Volume	R_V	
cm^3	Ci s Gy^{-1}	Gy m^2 s h^{-1} Gy^{-1}	cm^3	Ci s Gy^{-1}	Gy m^2 s h^{-1} Gy^{-1}
2	43.0	0.173	56	396.2	1.599
4	68.2	0.275	60	414.8	1.674
6	89.4	0.361	64	433.0	1.748
8	108.3	0.437	68	450.9	1.820
10	125.6	0.507	72	468.4	1.891
12	141.9	0.573	76	485.6	1.960
14	157.2	0.635	80	502.5	2.028
16	171.9	0.694	84	519.1	2.095
18	185.9	0.750	88	535.5	2.161
20	199.4	0.805	92	551.6	2.226
22	212.5	0.858	96	567.4	2.290
24	225.2	0.909	100	583.1	2.354
26	237.5	0.959	120	658.5	2.658
28	249.6	1.007	140	729.7	2.945
30	261.3	1.055	160	797.7	3.220
32	272.8	1.101	180	862.8	3.483
34	284.0	1.147	200	925.6	3.736
36	295.1	1.191	220	986.3	3.981
38	305.9	1.235	240	1045	4.219
40	316.6	1.278	260	1103	4.450
42	327.0	1.320	280	1158	4.676
44	337.3	1.362	300	1213	4.896
46	347.5	1.402	320	1266	5.111
48	357.5	1.443	340	1318	5.322
50	367.3	1.483	360	1370	5.528
52	377.1	1.522	380	1420	5.731
54	386.7	1.561	400	1469	5.931

time from hours to seconds. Table 6.3 presents a modified version of the Manchester volume implant table for HDR applications. The Wisconsin experience finds that most fairly regular implants agree to within 5%, and irregular implants to within 8%. If the prescribed dose varies in different parts of an implant, prorating the calculated time usually yields agreement also to within 8%.

Table 6.4. *Quality assurance index values for regular planar implants, derived from the original Manchester tables, corrected by replacing 1000 R with 9 Gy. Conversion to Ci s by using the factor* 1.79×10^{-3} *Ci mg Ra eq*$^{-1}$*, conversion to air kerma rate by using* 7.225×10^{-3} *Gy m*^{2} *h*$^{-1}$ *mg Ra eq*$^{-1}$.

R_A for HDR Planar Implants					
Corrected and specificed to the periphery					
Area	R_A		Area	R_A	
cm^2	Ci s Gy^{-1}	Gy m^2 s h^{-1} Gy^{-1}	cm^2	Ci s Gy^{-1}	Gy m^2 s h^{-1} Gy^{-1}
1	53.38	0.2155	52	569.1	2.298
2	76.13	0.3074	54	584.0	2.358
3	94.20	0.3803	56	598.2	2.415
4	110.6	0.4467	58	613.1	2.475
5	126.4	0.5104	60	628.0	2.536
6	138.9	0.5611	62	642.1	2.593
7	150.7	0.6087	64	657.0	2.653
8	161.7	0.6530	66	671.1	2.710
9	173.5	0.7005	68	685.3	2.767
10	184.4	0.7449	70	698.6	2.821
11	194.6	0.7861	72	713.1	2.878
12	204.9	0.8272	74	728.0	2.938
13	215.1	0.8684	76	742.2	2.995
14	226.0	0.9128	78	755.6	3.052
15	237.0	0.9571	80	769.8	3.109
16	247.2	0.9983	84	797.4	3.220
17	257.5	1.040	88	825.8	3.334
18	268.4	1.084	92	853.4	3.445
19	278.6	1.125	96	881.1	3.556
20	288.9	1.166	100	906.6	3.661
22	308.5	1.245	120	1026	4.142
24	327.3	1.321	140	1148	4.637
26	347.0	1.401	160	1262	5.097
28	365.8	1.477	180	1371	5.532
30	384.6	1.553	200	1476	5.960
32	402.7	1.626	220	1576	6.364
34	421.5	1.702	240	1674	6.756
36	438.0	1.769	260	1771	7.152
38	456.1	1.841	280	1862	7.516
40	473.4	1.911	300	1959	7.908
42	489.8	1.978	320	2058	8.308
44	505.5	2.041	340	2148	8.676
46	514.2	2.076	360	2239	9.041
48	537.7	2.171	380	2330	9.405
50	553.4	2.234	400	2418	9.761

Table 6.5. *Values for calculating* $\text{index}_{i,2}$ *in equation (6.14).*

h (cm)	a (cGy cm^2 Ci^{-1} s^{-1})	b (cGy cm Ci^{-1} s^{-1})	c (cGy Ci^{-1} s^{-1})
1.5	0.402	0.923	−0.0171
2.0	0.258	0.836	−0.0148
2.5	0.619	0.751	−0.0126
3.0	0.822	0.673	−0.0106

Two planar implants follow a similar line of reasoning as volume implants when using Manchester-type (i.e. optimized) distributions with Paris-type dose specifications, evaluated with the values in table 6.4. One-plane implants, however, pose different problems. The Paris system defines the basal dose as the average of the local minima between the needles in the implant plane and the reference dose as 85% of the basal dose. The Manchester system defines the dose on a plane 0.5 cm from the implant plane, and in that plane, approximately 10% above the dose at points across from the edge of the implant. The two systems also tend to use different needle geometries. However, despite the differences, table 6.4 often works well for verification of single plane implants using either (or no) system.

Ezzell (1994) derived an evaluation index particularly for optimized, rectangular, single-plane implants using HDR brachytherapy,

$$\text{index}_{i,1} = \frac{\text{dose}(h)[\text{cGy}] \times \text{area}[\text{cm}^2]}{\text{source strength [Ci]} \times \text{total time [s]}} \tag{6.13}$$

where $\text{dose}(h)$ = the dose prescribed to a plane a distance h cm from the implant plane. If E = the equivalent square of the implanted area (evaluated as 4 area/perimeter), a value used for evaluation of $\text{index}_{i,1}$ becomes

$$\text{index}_{i,2} = a + bE + cE^2 \tag{6.14}$$

where the values for a, b and c depend on the treatment distance h, and can be found in table 6.5. Ezzell suggests that $\text{index}_{i,1}$ should agree with $\text{index}_{i,2}$ to within 5% for planes without significant curvature.

6.4. WHEN THE CHECKS FAIL

Sometimes a correct treatment fails the consistence tests. The problem when faced with such failures becomes to assess whether the failure stems from some unusual aspect of the treatment or a mistake in the generation of the treatment plan. The first action to take is to verify that all of the assumptions for the consistency test pertain to the particular application. Unusual spread of ovoids or divergent template needles may produce indices outside those normally expected, but be perfectly appropriate for the patient.

Table 6.6. *Essential elements of the treatment prescription process. The individual listed first is responsible for executing the activity while the second individual is responsible for verifying the procedure. (5)*

Activity	Methodology
Radiation oncologist/physicist	
—Complete, sign and date prescription —Review radiographs and mark treatment and/or target volume and sign	—Physicist/dosimetrist: Question deviations from treatment policies or customary practice
Radiation oncologist and physicist	
—Define dose prescription and optimization criteria —Define treatment-planning constraints (maximum and minimum doses, normal tissue tolerances etc)	—If high frequency procedure or clinical indication for treatment, written treatment policies which describe the dose specification and optimization processes should be available —If procedure is infrequent, customized or otherwise unusual, physician and physicist should work together to address these questions
Physicist/dosimetrist	
Active dwell positions selected and documented	—If standard pattern used (e.g., intracavitary implant), procedure type and protocol identified —If identified and marked radiographically, confirm that all relevant data defining target volume are consistent with proposed dwell position distribution

Finding an abnormality may or may not explain the variance, but if it appears to and all parties agree that all other indications point to the treatment being correct, the treatment may proceed. If any one of the team members has hesitations, a second running of the plan (preferably on a different system, or by different operators as discussed above) should be initiated. The two plannings should produce dose distributions that agree to within 2% (or, in regions of gradients greater than 2% mm^{-1}, in position to within 1 mm) when superimposed.

Persistent discrepancies, as discussed in chapter 1, call for abortion of the procedure. Part of the purpose of the quality assurance tests is to prevent execution of errors. Continuation of the treatment procedure in the face of indications of errors renders the entire quality control process a waste of time and effort.

Table 6.7. *Critical steps in the treatment planning process. The individual listed first (primary) is responsible for executing the activity while the second individual (secondary) is responsible for verifying the procedure. (6)*

Activity	Methodology
Physicist/dosimetrist	
Review treatment planning procedure	Physicist reviews with dosimetrist: —which written procedure, if any, to be followed, or —identifies reconstruction, optimization, dose specification procedures to be used for this case
Dosimetrist/physicist	
Active dwell position localization	—If standard pattern used (e.g., intracavitary implant), procedure type and protocol identified —Channel numbers matched to radiographic image, treatment length and first and last dwell positions in each catheter calculated. Physicist to review
Dosimetrist/radiation oncologist	
Verify plan input data Assess clinical adequacy of plan Accept or reject plan	—Compare patient name on prescription, radiographs, localization data and HDR treatment schedule —Confirm date/time displayed on computer planning system, and that displayed source strength agrees with source inventory or chart —Check entered daily dose against prescription, for each catheter check length, dwell spacing and active dwell position numbers against localization protocol or planning procedure —Check radiograph orientations, distances, magnifications and gantry angles against requirements for selected source position reconstruction algorithm —Intended volume treated to desired dose —Optimization goals and constraints satisfied

6.5. SUMMARY

High dose rate brachytherapy requires rapid and sure gathering and transmission of information for the treatment plan. Minimizing error entry into this information makes extensive use of protocols, forms and, particularly, independent second persons monitoring the assembly and entry of data. Regardless of the care taken to prevent errors from entering the treatment plan, they occasionally will. To prevent the execution of erroneous treatments, the procedures must also contain steps to detect errors before treatment delivery. Error detection utilizes independent reviews.

The error-detection reviews should follow a set procedure using forms that check the following aspects of the generated plan:

(i) Dose prescription, including:

(1) comparison of the delivered dose to the prescribed dose,
(2) evaluation of the uniformity of the dose,
(3) location of the prescribed dose,
(4) dose distribution/differentials in dose,
(5) dose distribution begins and ends correctly along the catheter.

(ii) Dose to normal structures.
(iii) Programming, including:

(1) program identification,
(2) date and data file information,
(3) treatment length,
(4) step size,
(5) correct assignment.

Several 'indicators' for correctness or consistency are included in the chapter to assist in evaluating the plan.

Assuming that the checks of the plan indicate that everything is in order, before treatment the following should be checked at the treatment unit:

(i) Patient identification.
(ii) Program identification.
(iii) Program verification, including:

(1) the source strength compared to that on the treatment plan,
(2) the step size,
(3) the length to the first dwell position for each channel,
(4) each dwell position used and its respective dwell time and
(5) the total treatment time.

(iv) Channel verification.

APPENDIX 6A. SOME RECOMMENDATIONS FROM THE REPORT OF TASK GROUP 59 OF THE AMERICAN ASSOCIATION OF PHYSICISTS IN MEDICINE

Tables 6.6–6.10 are from the report of Task Group 59 of the AAPM (1998) and provide some suggestions for consideration in establishing a high dose rate programme. Addressing the issues discussed carries quality planning through many of the important phases of the processes necessary for quality treatments. The table number in the report follows the caption in parentheses. All the tables are used with the permission of the American Association of Physicists in Medicine.

Table 6.8. *Pretreatment physicist review of HDR treatment plan and dwell-time calculations. (7)*

Endpoint	Check methodology
Patient identity	Compare patient names/numbers/dates printed on prescription, simulator radiographs, chart and localization form
Input data	As described in text
Positional accuracy/ implant geometry	—Applicators modelled in treatment plan match those of operating room description and implant diagram —Verify matching and localization calculations against radiographs if interstitial or transluminal implant —Compare active dwell positions, dwell separation and treatment length listed on computer plan to localization form or to appropriate treatment planning procedure —Compare three orthogonal dimensions of implant measured from AP and lateral radiographs to corresponding dimensions of graphic plan —Check radiograph orientations, distances, magnifications and gantry angles against requirements for selected source position reconstruction algorithm
Plan optimization process	—Appropriate optimization option used —Dose optimization and dose specification points in correct location relative to dwell positions on graphic plan —Expected isodose curve passes through dose specification points —Optimization algorithm produces expected distribution of dwell weights, coverage of target volume and distribution/magnitude of hot spots or peripheral/central minimum dose ratio. Implant quality parameters derived from dose–volume histograms, if available and previously validated, should be checked
Dose calculation accuracy	—(IRAK)/dose ratio falls within expected range —Assuming distribution of dwell times on computer plan printout, manually calculated dose agrees with dose calculated by computer planning system within expected tolerance —Doses at clinically important points of interest agree with values interpolated from isodoses —Isodose curves calculated in appropriate planes
Clinical adequacy	—Prescribed dose, applicator selected and dose distribution consistent with policies of treatment for patient's disease or physicist's understanding of physician's clinical intent —Volume covered by prescription isodose surface consistent with all known target localization data —Maximum dose and dose to critical anatomic structures, including previously administered therapy, within accepted range
Daily treatment record	Source strength, total dwell time, total IRAK, No and type of applicators correctly entered into daily treatment record

Table 6.9. *Pretreatment safety checks to be performed after patient setup and treatment programming. (11)*

Endpoint (Individual)	Methodology
Patient identity (Operator)	Any method appropriate to patient mental condition and degree of operator familiarity with patient. Check patient name against that on chart and treatment plan
Prescription followed (Operator)	—Compare dose/fraction on treatment plan and prescription —Compare total dose and No of fractions against treatment previously given —Compare source strength on treatment plan and remote afterloader —Check: prescription filled out and signed, physician and physicist have signed off on treatment documentation —Correct length transfer tubes used; applicator No–channel No correspondence correct —Dwell times and positional settings printed by afterloader match treatment plan
Setup accuracy (Physicist/operator)	—Programmed afterloader settings match treatment plan —Correct length transfer tubes used; applicator No–channel No correspondence correct —Applicator positioning checked against operating room measurements, reference marks

Table 6.10. *Post-treatment checks. (12)*

Endpoint (Individual)	Methodology
Patient/personnel safety (Operator)	—Area monitor checked before entering room. Enter room leading with survey instrument to confirm complete retraction of source —HDR device shut down and secured after patient removed
Treatment accuracy (Operator)	—Fill in daily treatment record —Compare total dwell time on treatment unit printout agrees with calculation
Chart order (Operator)	—All forms and checklists complete and properly filled in chart

CHAPTER 7

QUALITY MANAGEMENT FOR LOW AND MEDIUM DOSE RATE REMOTE AFTERLOADERS

For the most part the advantage of low dose rate remote afterloaders only comes from the reduction of the radiation exposure to personnel. Some of the units facilitate some increased optimization capabilities compared to manual low dose rate applications, and in some cases use the same sources as with manual loading. The reduction in personnel exposure led to wide use of remote afterloaders throughout Europe, although it never seemed important enough to overcome the price of the units to produce a large market in the United States. While high dose rate brachytherapy units tend to follow very similar designs (however, with important differences in features), the LDR units vary remarkably, and quality considerations almost must address each available unit.

The title for this chapter includes low dose rate and middle dose rate treatment units because the middle dose rate machines intend to function as LDR units, except with particularly strong sources. That difference in source strength, however, leads to biological consequences potentially dangerous to the patient.

Terminology fails for some of the differentiations necessary for the discussion of LDR remote afterloader. For the present purpose, the following definitions prove useful, but the reader should recognize that these terms convey no distinctions outside this text, and as applied to some units, the designations may belie their complexity.

Simple source loader. A machine that accepts low dose rate brachytherapy sources identical to those used for manually loaded cases, and simply moves them into the appliance in place in the patient. These units reduce (but often do not eliminate) the exposure to persons preparing the sources and inserting them into the patient. Exposure to hospital personnel attending the patient and visitors remain unchanged since these units leave the sources in place over the duration of the treatment.

Basic remote afterloader (fixed source train afterloader). A machine that uses low dose rate sources similar, but not necessarily identical to, those used for manually loaded cases, that operates with all personnel outside the patient's room, moving the sources into the appliance. These units move the sources into a shielded container before personnel enter the room to attend the patient.

Advanced remote afterloader. A machine that uses sources markedly different from manually loaded LDR sources, that allows for some optimization of the dose distribution, with the same protection characteristics as a basic remote afterloader.

As with the HDR remote afterloaders, the main concerns for LDR afterloaders centre around accurate positioning of the sources for the correct time, the strength of the sources and patient safety. Many of the checks closely parallel those for the HDR units, with only the differences for these tests highlighted in this chapter.

In the discussion, the checks fall into those needed before every patient, those performed on a quarterly basis and those required annually. The designation 'quarterly' here implies only a suggestion based on users' recommendations. Part of the job for the quality manager at each facility consists of determining the frequency for such periodic tests. The performance for some units may indicate a shorter period for some of the tests, for example monthly. If the period remains quarterly, that time interval would be expected to have slight variations around 122 days. Nothing particularly happens if the interval extends to 125 days on occasion. However, to be effective, the gating of the periodic tests should not fall far from the targeted time.

The discussion below owes much to the work of Williamson (1991), Slessinger (1990a, b, 1995a, b) and Glasgow (1995), and particularly to conversations with Jeffrey Williamson.

7.1. ADVANCED REMOTE AFTERLOADERS

Some of the older advanced remote afterloaders use ^{60}Co pellets. However, most of the units currently in the field use spherical sources of ^{137}Cs of 2.5 mm diameter, transported between a storage safe and the treatment appliance pneumatically through attached catheters. This chapter will address only these latter units. For the caesium units, a treatment source train consists of a series of spheres that move through the catheter together into the desired position. During programming of the train, the operator selects either an active sphere or an inactive sphere to occupy each position in the train. While not as flexible as optimization with an HDR unit or manually loaded LDR iridium sources, such a procedure allows customization of the dose distribution through the selection process. During creation of the source train for a channel of the treatment unit, the source and spacer pellets reside in a shielded, temporary holding container. At loading, the operator connects the unit to the treatment appliance, leaves the

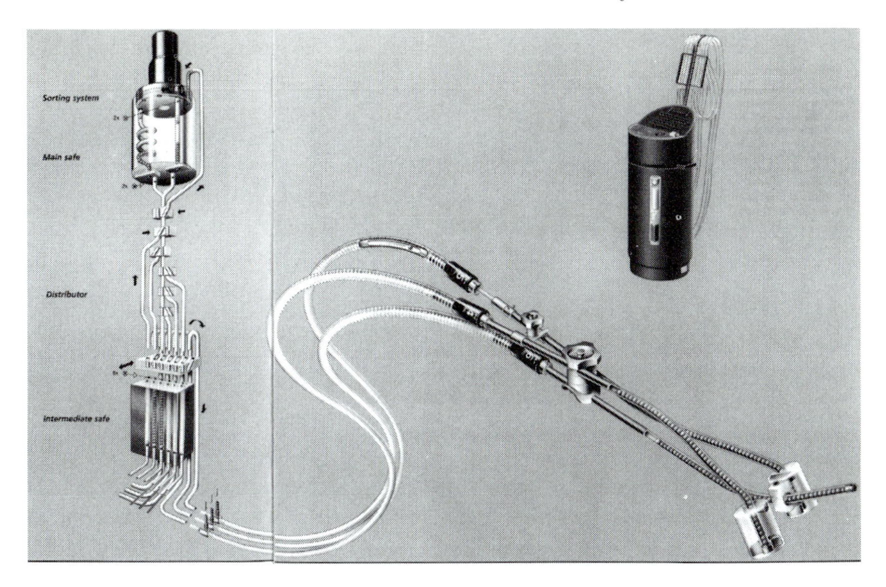

Figure 7.1. *The Selectron LDR, and advanced remote afterloader. This unit uses small active and inactive spheres pneumatically moved through tubes to create the desired shape for the dose rate distribution. The diagram shows a schematic of the important parts of the unit and the flow of sources. (Photograph courtesy of Nucletron Corporation, Columbia MD, USA and Veenendaal, The Netherlands.)*

room and instructs the machine to move the sources into position. The source train always moves to the end of the tube in the appliance where a control pellet blocks the airflow valve at the tip. The unit uses the halt in the airflow to register that the source train reached its destination. Figure 7.1 illustrates such a unit.

Personnel attending the patient press an interrupt button and the sources move temporarily into the shielded housing. When they leave the room, depressing the start button resumes the treatment. A timer in the unit clocks the cumulative time the sources reside in the treatment position. At the end of the treatment duration, the unit withdraws the sources automatically. Magnetically, the unit sorts the nonmagnetic active pellets to the shielded container and the magnetic nonactive pellets to a different storage tube. The size of the sources limits the use of these afterloaders to intracavitary applications.

7.1.1. Checks per patient

The following constitute checks performed prior to each use. Much of this list duplicates the tests for an HDR unit. Note that most of these tests require performance before the patient enters the room, possibly during the placement of the appliance.

7.1.1.1. Communication equipment. See that the television and intercom systems function. PASS: television images of the room are visible and sounds in the room can be heard; FAILURE: either communication device fails to operate.

7.1.1.2. Date and time. Check that the computer contains the correct time and date information in the required format. PASS: the time and date are correct and in the correct format; FAILURE: either the values or format for the date or time differ from that expected.

7.1.1.3. Source information. If the unit uses stored information about the sources for the determination of the actual treatment time, such as their average strength corrected for decay, ensure that that information is current and correct. PASS: the strength equals that predicted by correcting the initial strength for decay within the rounding uncertainty of the data precision; FAILURE: the difference between the source strength in the treatment unit exceeds rounding uncertainties (usually greater than about 0.2%).

7.1.1.4. Drive air pressure. For units that use air pressure to drive the sources to the treatment position, check that the pressure falls in the operating range. PASS: the pressure falls within the range; FAILURE: the pressure falls below *or* above the operating recommendations.

7.1.1.5. Applicator attachment. Program the unit to send a source to a location in each channel without attaching an appliance. PASS: the unit refuses to send the source out any of the channels; FAILURE: the sources exit the unit.

7.1.1.6. Catheter attachment lock. Attach catheters to each of the channels in use but do not lock them in place, and try to initiate source movement. PASS: the unit refuses to pass a source into any of the catheters; FAILURE: the source exits one or more of the channels.

7.1.1.7. Pathway patency. Remove all the catheters from the channels, and attach and lock to one channel a special catheter with a partial blockage in the middle. Program the source to travel to the end of the catheter and initiate a run. The unit should detect the failure of the source to seat in the proper position. PASS: the unit refuses to send out the source; FAILURE: the unit sends the source into the catheter.

7.1.1.8. Program completion. Replace the blocked catheter with a patent one, and program the unit with some missing information, such as a source train configuration but with no time. PASS: the unit will not initiate source movement; FAILURE: the source passes into the catheter.

7.1.1.9. Door interlock. Program the source train for a short time, such as 1 min. With the door to the room open, try to initiate a run. PASS: the unit refuses to advance the source into the catheter; FAILURE: the unit sends the source into the catheter.

7.1.1.10. Warning lamps. Close the door, and initiate the source run. Observe the warning lamps by the door. PASS: the lamp comes on when the sources pass into the catheter; FAILURE: the lamp fails to light.

7.1.1.11. Room monitor operation. With the sources in the catheter from the previous test, open the door and observe the proper operation of the room monitor. PASS: the monitor signals the presence of radiation; FAILURE: monitor fails to indicate the presence of radiation in the room.

7.1.1.12. Hand-held monitor. While performing the previous test, hold the handheld monitor in the doorway to see whether it indicates the presence of radiation. PASS: the hand-held monitor indicates a radiation reading; FAILURE: the monitor fails to respond.

7.1.1.13. Door interrupt. During the previous two tests, the unit should have been retracting the sources. PASS: the unit retracts the sources; FAILURE: the exposure continues with the door open.

7.1.1.14. Treatment interrupt. Close the door and reinitiate the exposure. Once the sources reach the treatment position, press the 'treatment interrupt' button. PASS: the unit retracts the sources; FAILURE: the sources remain out of the shielded housing.

7.1.1.15. Emergency stop. Restart the unit. Once the sources reach the treatment position, press the 'emergency off' button. PASS: the unit retracts the sources; FAILURE: the sources remain out of the shielded housing.

7.1.1.16. Timer termination. Reinitiate the exposure and let it continue until the elapsed duration equals the time set on the timer. PASS: the unit stops the exposure and retracts the sources; FAILURE: the exposure continues past the time for termination.

7.1.1.17. Appliance and transfer tube condition. Check the condition of the appliance and tubing for use with the patient. PASS: the appliance and tubing appear in good condition with connectors clean and clear; FAILURE: any part of the appliance, tubing or connector shows wear, breaks, blockages or dirt. (If this set of checks coincides with the placement of the appliance, obviously the check of the appliance proper must have been preformed previously. For a more detailed discussion of quality assurance procedures for applicators, see chapter 8.)

7.1.1.18. Computer supplies. Check that the supply of paper and ink (or equivalent) will last through the treatments. Some units will initiate a source retraction if they detect the inability to record the progress of the treatment. PASS: supply is adequate to allow treatment; FAILURE: the supply may not last through the treatment.

7.1.1.19. Autoradiography. Program the source trains to be used in the patient. Place the catheters on a sheet of film in a paper jacket. Slide localization markers into the catheters and mark on the film beside the marker locations that correspond to active pellets in the treatment. Initiate an exposure of approximately 6 min divided by the source air kerma strength in U. Process the film and check that the active pellets went to the indicated locations. (Details on autoradiography for a remote afterloader can be found in the chapter on quality assurance for high dose rate treatment units. Exposure of the sources during the autoradiography using a diagnostic radiography unit helps locate the source positions more clearly than the simple marks on the film, but usually requires coordination of mobile x-ray units into the patient's room during the time of the testing, an unsure task in many facilities.)

7.1.2. Periodic checks

Periodic checks for these afterloaders include all of the checks per patient plus those below. Note that the autoradiography test from the per patient differs slightly for the periodic tests as discussed below.

7.1.2.1. Quarterly tests

Pressure sensor. The method for testing the drive pressure sensing system, as well as the actual frequency for this test, follow the manufacturer's suggestion.

Source identity (autoradiography). Program a standard configuration that uses all of the active pellets, and make an autoradiograph as in section 7.1.1. PASS: each of the expected active pellets produces a darkened image on the film; FAILURE: the autoradiograph indicates that active and nonactive pellets have been confused.

Timer accuracy and linearity. Set a treatment time of approximately 2 min, such that the travel time for the sources becomes a negligible fraction of the total time. Initiate the exposure and clock the elapsed time with a stopwatch. Repeat the measurement using a longer time or approximately 15 min. PASS: for both exposures the measured times and the set times should agree to within approximately 2 s, with the difference being an additive constant, indicative of the travel time. The measured times might be longer or shorter than the set times depending on whether the unit begins the clock when the source train leaves the unit or settle into treatment position; FAILURE: any measured time differs from the set time by more than 4 s, or one measured time is longer than that set and the other is shorter.

Power backup. Program the unit and initiate an exposure. During the exposure remove power for the unit by opening the circuit breaker or unplugging the unit if that is possible. PASS: the system's backup power supply takes over operation (the unit may continue to treat—recording the time on the cumulative clock and operating any interrupts—or retract the sources depending on the design intention); FAILURE: the sources remain out but the clock stops or the interrupt fails to operate by the button or opening the door.

X-ray marker. Accurate calculation of dose distributions depends on the x-ray markers correctly simulating the positions the sources will occupy. Problems arise with the markers due to stretching or kinking of the carrier (depending on the construction), shifting of the location of the actual marker beads along the cable that carries them, or failure of the marker cable to properly seat in the applicator tube. Testing the integrity of the markers requires two steps.

(1) Measure compared to a tape measure the positions of the marker beads compared to their intended location. PASS: the beads all fall within 1 mm of their intended locations; FAILURE: one or more beads fall out of the correct location. (Of particular concern is when the seeds progressively fall further from their correct location along the cable.)

(2) Insert the marker cable into an applicator tube. Place a small, radiopaque object, such as lead shot, on the tip of the tube. Radiograph the tube and compare the distance between the simulated control pellet and the end of the tube with the manufacturer's specification of that distance. PASS: the control pellet falls at the correct distance from the end of the tube; FAILURE: the control pellet falls more than 1 mm farther from the end of the tube than specifications.

Detached source indicator. Run the sources into a treatment catheter and disconnect the catheter from the unit. PASS: the unit immediately alarms indicating the disconnection (or prevents the disconnection in the first place); FAILURE: the system makes no indication of the detachment.

Integrity of connectors. Test connections for all transfer tubes and appliances to assure they connect solidly and only between parts intended. PASS: all connectors function properly and cleanly; FAILURE: connectors fail to couple parts as intended, make loose connections, make connections only with unusual force or allow connection of parts (or parts to channels) that should not connect.

7.1.2.2. Annual tests. The annual tests include the quarterly tests, plus the additional tests below.

Source strength determination. Calibration of spherical sources for such afterloaders usually follows two steps. First, an ionization chamber with suitable buildup and a standard, calibrated ^{137}Cs source inside a plastic catheter are positioned in a low scatter environment with at least 20 cm between, and the source oriented to place the chamber on its perpendicular bisector. A reading taken in this geometry provides the chamber with a calibration factor for ^{137}Cs, assuming that the chamber has not received such a calibration at a standards laboratory. The catheter containing the source is then replaced with one for the afterloader, and a train of sources comprising the same active length is then moved into the catheter in the same geometry as the original ^{137}Cs source. This measurement gives the strength of the train, S_t, by simple proportion to that of the standard source. Each of the spheres in the train is then measured in a well chamber, giving a reading r_i, to establish their relative strengths, and their absolute strength deduced from

$$s_i = S_t \frac{r_i}{\sum_j r_j}$$

where the sum equals the total of all the readings. The strengths s_1 for any of the rest of the sources in the unit would then come from its reading, r_1, and the equation

$$s_1 = r_1 \frac{s_i}{r_i}$$

where the fraction simply represents the proportion between the strength and reading for one of the now-calibrated sources.

Source integrity tests. Test the sources for leaks either as per discussion in the chapter on low dose rate sources if the system allows access to the sources, or by wiping the channel openings and the inside of a used catheter if the unit prevents access to the sources. PASS: no activity is detected on the wipe sample; FAILURE: the wipe test procedure indicates removable activity on the sample. NOTE: some governments require this test more often than annually.

7.2. BASIC REMOTE AFTERLOADERS

Basic remote afterloaders include devices with fixed source trains. These devices contain a bank of sources and/or source trains already in commonly used configu-

Figure 7.2. *The microSelectron LDR, an example of a basic remote afterloader. (Photograph courtesy of Nucletron Corporation, Columbia, MD, USA and Veenendaal, The Netherlands.)*

rations. The sources move from the storage compartment to the patient by use of a drive cable that latches on to the source train and pushes it down the connecting tube and into place in the treatment appliance.

The basic remote afterloaders may use either ^{192}Ir or ^{137}Cs (at the time of writing). The iridium source trains are the same as would be used in the patient were the case manually afterloaded. The operator attaches a special coupler to the train and places it in the desired channel of the unit's storage safe. If the ribbons in the channels differ or contain asymmetric seed distributions, the operator must take great care to load the sources into the unit correctly. As with standard manual afterloading, a second person should watch the loading to verify that it matches the treatment plan. The caesium sources form a permanent inventory and live in the shielded storage safe in the unit. Since they serve for intracavitary sources, the ^{137}Cs trains usually correspond to typical loading for LDR cervical applications, for example, with a strength in the distal-most 2 cm equivalent to 15 or 20 mg of radium and the equivalent of 10 mg of radium in the next 2 cm and also the 2 cm following that. The inventory also would contain the equivalent of single source 'trains' for ovoids. During assembly of the sources for a particular patient, the unit moves the designated sources from the storage bank into the interim safe. During interruptions of the treatment for patient care, the sources move back into the interim safe.

While not using pneumatic pressure to move the sources, one of the more common units uses airflow as an indicator that the sources arrived at the end of the catheter. Air flows down the tube until a control pellet at the end of the source train connector abuts proximal (machine) end of the appliance and blocks the flow.

Programming the unit may require determining several specified lengths, as discussed by Williamson (1991). Any trimming of catheters following these measurements requires re-establishing the length parameters.

Figure 7.2 shows one such unit.

7.2.1. Checks per patient

Note again that most of these tests require performance before the patient enters the room, possibly during the placement of the appliance. Many of these tests mirror those for the advanced remote afterloader. The list below either follows the same procedure as above, or differs minimally in varying ways depending on the particular unit.

Communication equipment.
Date and time.
Source information.
Drive air pressure.
Applicator attachment.
Catheter attachment lock.
Pathway patency.
Program completion.
Door interlock.
Warning lamps.
Room monitor operation.
Hand-held monitor.
Door interrupt.
Treatment interrupt.
Emergency stop.
Timer termination.
Computer supplies.
Appliance and transfer tube condition.

While the following tests apply to both the advanced and the basic afterloaders, some differences become important to consider:

7.2.1.1. Connector condition. Check the condition of both sides of the connector that couples the source train to the drive cable. PASS: the connectors appear in good condition, clean and without bends or defects; FAILURE: any part of the connector shows wear, breaks, blockages or dirt. NOTE: uncoupling of the connection between the source and drive cable has been a frequent event with many afterloading devices of this nature. Quality control procedures probably will not eliminate the problem, but may reduce the frequency.

7.2.1.2. Autoradiography. This test becomes extremely important for these units, particularly those for which the storage facility contains both ^{192}Ir and ^{137}Cs. Since the operator manually connects the treatment channels to the storage locations during 'programming', no feature in the unit checks that each channel picks up the correct source train. Thus, evaluating that each channel contains the correct source configuration becomes an even more important check than with the advanced units. Performance of this test follows the general guidance discussed previously. As discussed in the chapter on high dose rate brachytherapy, using markers compatible with making the x-ray image with the transfer tubes connected to the catheters also checks for errors in matching markers' images and specified treatment lengths. Remove the markers and connect the programmed channels. Move the sources into position for a time adequate to darken the film directly under the sources. For low strength ^{192}Ir sources, localization film probably provides the images in a reasonable time. The higher strength ^{137}Cs applications may produce cleaner images on verification film. PASS: upon processing the film, the centres of the images of the markers and the centres of the respective sources they represent match within 1 mm; FAILURE: the centres differ by more than 2 mm. NOTE: differences between 1 and 2 mm require further investigation. A failure in this test does not differentiate between a problem with the sources or programming, and a problem with the x-ray markers. The first step following a failure is to measure the length and positions of the marker beads along the marker train, and compare the measured values with those in the machine specifications.

7.2.2. Periodic checks

Periodic checks for these afterloaders include all of the checks per patient plus those below. Note that the autoradiography test from section 7.2.1 differs slightly from that for the periodic tests as discussed below.

7.2.2.1. Quarterly tests.

Pressure sensor. When appropriate, the method for testing the pressure sensing system, as well as the actual frequency for this test follow the manufacturer's suggestion.

Source identity (autoradiography). If the unit contains a permanent inventory of sources, program each of the source trains into an autoradiograph test tool or a catheter taped on a piece of film. Create an autoradiograph as described previously. PASS: each of the source trains produce a darkened image on the film with the expected geometry; FAILURE: the autoradiograph indicates active lengths or relative loadings different from expectations.

Detached source indicator. Connect a dummy ribbon to the drive for a channel. Program the unit to extend the dummy train for about two minutes, and run the ribbon into a catheter with most of the end cut away. Once the train comes into position, detach the connector from the drive cable. PASS: when the drive cable retracts, the unit alarms the indication of a detached source; FAILURE: the unit issues no notification of a problem with the retracted source.

Partial source loss. Reset the unit and reconnect the dummy ribbon. Again send it into the cutaway catheter. Cut off approximately 1 cm of the ribbon. PASS: when the train retracts, the unit alarms the indication for a shortened source; FAILURE: the unit issues no notification of a problem with the retracted source.

The tests below follow the same procedure as with the advanced remote afterloaders.

Timer accuracy and linearity.
Power backup.
Integrity of connectors.

7.2.2.2. Annual tests. The annual tests include the quarterly tests, plus the additional tests below.

Source strength determination. Because some of the sources form permanent source trains of unusual length or nonuniform distribution, normal assay procedures often fail to yield the correct source strength and certainly tell nothing of the linear weighting of the strength. One approach to this problem follows three steps.

(1) The first step entails establishing the strength of one of the sources with a 'usual' active length, approximately 2 cm. For the most part, the sources under discussion usually are ^{137}Cs sources, however, with a different source construction from the manually loaded sources. As such, few well-type chambers carry calibrations for these sources. The assay of a remote afterloader's source comes from comparison with a manual source assayed as in the chapter on low dose rate sources.

   (a) In a low scatter geometry (discussed in chapter 2) set the assayed source approximately 20 cm from an ionization chamber with an appropriate buildup cap (at least 3 mm for ^{137}Cs). Take a reading R_{cal}.
   (b) Replace the assayed source with the 2 cm source from the remote afterloader, henceforth called the 'standard' for this unit. If necessary for nonremovable sources, run the source into a catheter. Take a reading R_{std}.
   (c) Calculate the strength of the standard using the usual proportionality formula:

$$S_{std} = S_{cal}\frac{R_{std}}{R_{cal}}. \tag{7.1}$$

(2) Build a shielded collimation geometry as described by Slessinger (1989), reproduced in figure 7.3. The 5 cm blocks provide a reduction in the radiation passing through to approximately 0.3%. This geometry attenuates greatly the radiation incident on the detector from all but 2 cm of the source. Measure the reading produced by the standard afterloading source in this comparison geometry, ${}_{comp}R_{std}$.

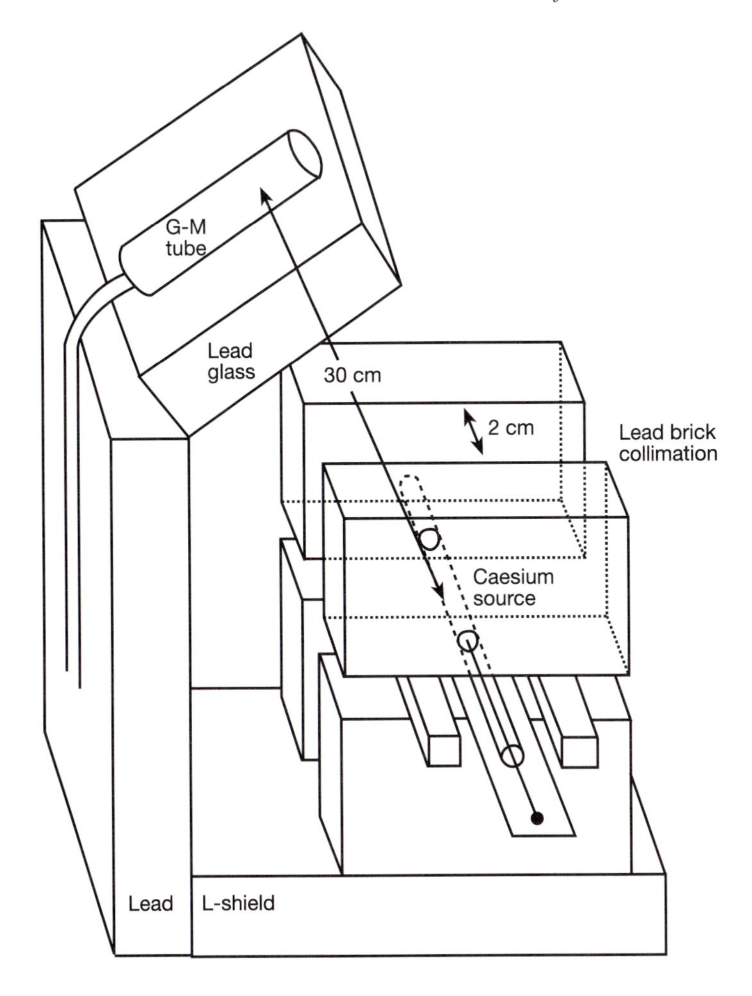

Figure 7.3. *A method for establishing the strengths of sources in a source train. (Reproduced from Slessinger 1989, 1995a by permission.)*

(3) Step the afterloading sources and trains through the assay window, 2 cm at a time, taking a reading ${}_{comp}R_{seg}$. Again using the proportional response formula, the strength of each 2 cm segment, S_{seg}, becomes

$$S_{seg} = S_{std}\frac{{}_{comp}R_{seg}}{{}_{comp}R_{std}}. \tag{7.2}$$

The transitive steps are required because the source construction differences could perturb the comparison between the train's segments and the calibrated source in the collimated geometry. For any other of the afterloader's 2 cm sources, assay can use the standard source to determine the proportionality constant for a well-type chamber, since the construction geometries remain constant.

Source integrity tests. If the unit allows removal of the sources, wipe the sources as discussed in the chapter on low dose rate sources, otherwise as described previously for the advanced afterloaders.

7.2.3. Procedural quality considerations with basic remote afterloaders

Aside from the unit quality control performed on the unit, for each application in a patient four additional concerns require addressing.

7.2.3.1. Connection confusion. During connections between the unit and the appliance in the patient, each channel's transfer tube should be connected completely from the unit to the catheter or needle in the patient one at a time to minimize the possible mixing or cross connection among the channels.

7.2.3.2. Appliance movement. With any low dose rate brachytherapy application, the patient moves during the treatment, potentially dislodging or mispositioning the appliance. While some interstitial applications provide relatively secure anchoring, intracavitary cases seldom encounter that luxury. In addition to the normal applicator movement seen in LDR cases, for those treated with a remote afterloader, the transfer tubes act as tethers that can produce additional forces on the appliance. In the wrong geometry and with the wrong length, patient movement pulls against the unit and displaces the applicator. Even in normal operation, the tubes present encumbrances to the patient and the risk of pulling on the applicator with unconscious leg movements. To guard against the delivery of a significant portion of the dose with inappropriate placement of the appliance, the nursing staff should be instructed in the appearance of the applicator in the patient, check the application frequently and contact the physician if the applicator looks different from its initial placement. Further discussion of related concerns follow below in section 7.2.4.

7.2.3.3. Treatment termination. At the end of treatment, the unit should withdraw the sources from the patient into the shielded container. As a backup, in case of machine failure, a trained person should check for source retraction shortly after the expected termination time. Interruptions in the treatment for patient care prevent calculation of the exact termination time, but, using experience as a guide, an operator can approximate the time, and ensure termination as of the elapsed time on the controls. With appropriate training in recognizing treatment beyond the set duration, the nursing staff on the unit provides a cadre to perform this check. They need training to recognize indications of unshielded sources and in operation of detectors and the unit interrupt to perform their routine functions; the additional training simply extends slightly their training syllabus. Alternatively, one of the radiation oncology staff can be present at the projected termination time. Unlike high dose rate treatments, failure of the unit to retract the sources does not constitute an emergency situation. Sources remaining in the patient beyond the treatment time normally exceed the prescribed dose by about 1% to 2% per hour. (Radiation levels in the room only slightly exceed that normally encountered attending a manually afterloaded patient, so the staff rectifying the problem need not fear excess exposure.) Yet, quality patient care dictates ensuring termination after delivery of a dose as close to that prescribed as possible.

7.2.3.4. Source removal. As with *any* brachytherapy procedure, following source retraction, a survey of the patient with a nonparalysable Geiger counter ensures that no source remained behind in the patient for unexpected and unexplained reasons.

7.2.4. Reported operational problems with LDR remote afterloaders

Slessinger (1995b) reported the following problems encountered using a fixed source train based unit:

- source trains hanging on slight discontinuities on the inner wall of the catheter or at the connection between the source train and the drive wire;
- detachment of the source train from the drive cable in curved 'flexiguide' needles in template applications[1];
- source trains failing to seat in the proper location at the end of curved 'flexiguide' needles[1];
- patients dislodging the appliance by pulling against the catheters attached to the treatment unit.

Addressing this last problem, Williamson (1991) suggests great care in securing the appliance to the patient, and educating the patient to the importance of not pulling against the catheters. The experience of the author suggests that such problems happen most frequently while the patient either sleeps or is under heavy sedation and is not aware of their movements or the consequences. We have had a patient push hard enough with her heels away from a remote afterloader to tear loose a template sutured to her perineum while heavily sedated. Securing intracavitary devices proves more difficult than interstitial templates. While allowing adequate catheter lengths to accommodate normal movement and more, little can be done to completely prevent such patient related problems. An attentive nursing staff that notices when such problems occur prevents untoward sequeli in the aftermath. However, the environment of reducing health care costs by reducing nursing staff renders much less likely detection of such problems before rounds by the physician.

[1] Williamson notes that for these problems the treatment unit's alarm sounded, alerting the operator and allowing correction of the problem.

The first problem underlines the importance of acceptance testing all treatment appliances for *smooth* and proper operation. Williamson describes a solution to overcome the problem of discontinuities between the source and drive cable (covering the source and junction with woven-wire sheath and holding the source in the intermediate storage container rather than normal storage), but such problems constitute design or construction errors and should be solved by the manufacturer.

7.3. SIMPLE SOURCE LOADERS

The simple source loaders serve to minimize the time spent loading sources into treatment appliances. The operator loads the sources into the unit, transports the unit to the patient's room and connects the unit to the appliance using transfer tubes (ensuring that the proper channels connect to the correct parts of the appliance). On command, the unit moves the sources into the appliance, and the operator then disconnects the unit from the patient. Some of the basic remote afterloaders allow use in this mode. Since these units perform nothing with respect to timing, the quality control procedures reduce to assuring proper placement of the source trains. Proper placement entails moving each source train from the storage container into the correct needle or catheter, and inserting the train completely to the end of the track. There are no periodic tests, since all functions require evaluation per patient.

Autoradiograph. The program usually consists, as with the simple remote afterloaders, of the operator connecting the channel of the unit to source trains in the storage facility. Connection to the wrong storage location can produce treatment errors if source trains differ. The autoradiograph serves as the only check that the proper sources will occupy the correct locations. While not suitable for qualitative assay of individual sources, the autoradiograph displays significant differences in source strength as different darkening on the film, providing some indication of train direction for differentially loaded ribbons (that is, assurance that the operator did not load the train backwards). Executed as with the basic remote afterloaders, the autoradiograph also verifies the positioning of the train within the needle or catheter.

In other ways, such as assay of the sources, the simple source loader follows the procedures for manually loaded cases.

Figure 7.4 shows an example of a simple afterloader.

7.4. DOSE RATE CONSIDERATIONS

A fine line separates low dose rate and middle dose rate units. The important conceptual difference arises because the biological effectiveness of a dose delivered at rates called 'low' remains relatively constant regardless of rate. The same holds true at 'high' dose rates. However, the effectiveness at the low and the high dose

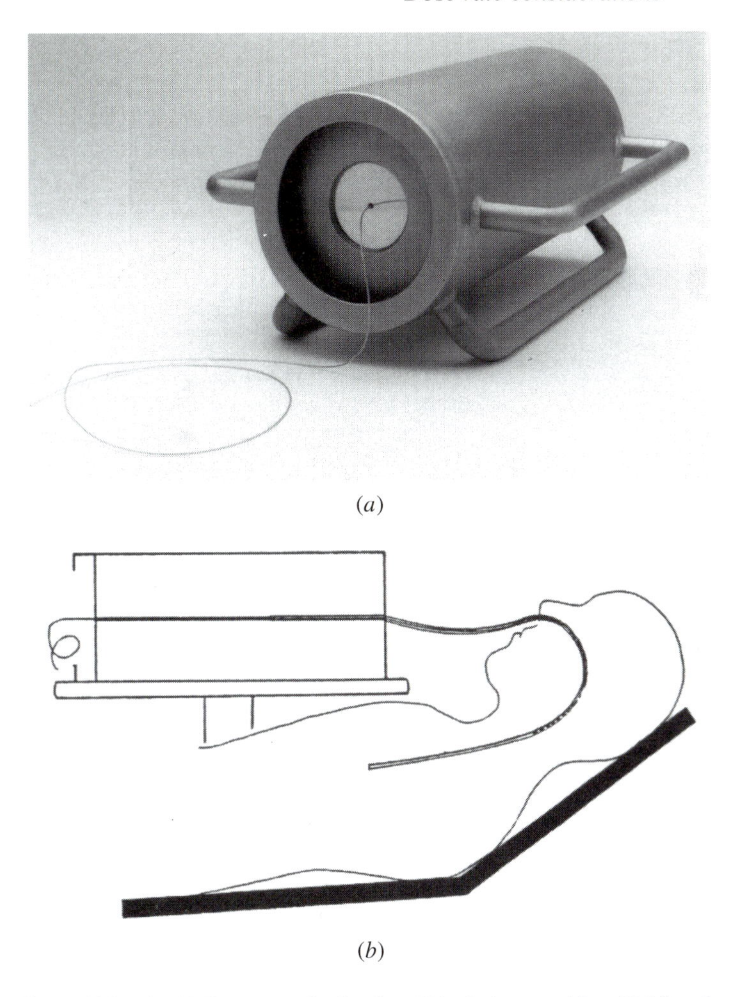

(*a*)

(*b*)

Figure 7.4. *The ALK remote afterloader. This device provides shielding for transporting sources from storage to a patient's room, and facilitates loading the sources into the treatment appliance. The device (a) has an interchangeable core allowing for a single strand (for treatments such as endobronchial insertions), multiple trains (for implants) and three larger channels for gynaecological intracavitary insertions. (Figures courtesy of Medical Radiation Devices, Inc., Dothan, AL, USA.)*

rates differ, the transition occurring in the large region between, called 'middle' dose rate. The effectiveness changes slowly at the boundaries, but low dose rate treatments generally cover the range from around 0.25 to 1 Gy h^{-1}. High dose rate treatments deliver the dose in times short compared to the half time of repair for sublethal cellular damage, or treatment times of less than 0.5 h. Several of the remote afterloaders deliver doses at rates above 1 Gy h^{-1}, but still over many hours or days, falling into the broad middle region.

Treatments falling in the middle dose rate range present the challenge of prescribing the proper dose when the biological effectiveness of the dose changes not only with the absolute dose rate (and, thus, varies throughout the target volume), but also with the dose itself. The biological effectiveness of a dose follows equation (6.4). Because typical values for α/β fall around 3 for normal tissues and 10 for malignant, the difference in biological effectiveness between the two depends on the dose rate. The sensitivity of normal tissue compared to tumour increases as the dose rate increases.

Since the dose rates usually fall into the lower part of the middle dose rate range (1.4 to 1.8 Gy h^{-1}), practitioners often simply use a multiplicative constant for the dose to adjust for the increased effectiveness. Hunter suggests decreasing the prescribed dose by 12 to 20%, noting that '... it seems unlikely that the clinical dose rate correction factors will ever be firmed up further...' (Hunter 1995). Quality control on the prescribed dose includes checking that the value contains corrections for the dose rate compared to that appropriate for a low dose rate application. As an evaluation, the multiplicative-factor approach probably serves adequately.

7.5. SUMMARY

Remote afterloaders for low dose rate brachytherapy fall into three categories: simple source loaders, basic remote afterloaders and advanced remote afterloaders. Each follows slightly different sets of quality control procedures, but the tests mostly fall into checking that the sources will occupy the correct locations in the patient for the proper time (if the unit includes timing), and that the sources move smoothly and surely through the various adapters and connectors. The sources for some of these units require fairly involved assay procedures because their geometry differs from that usually used in calibrating well-type ionization chambers.

Several potential situations frequently encountered during treatments with LDR remote afterloaders require consideration, including preventing confusion of source trains during programming or loading the sources into the unit, providing stability in the face of patient movement and assuring source unloading at the termination of treatment. Unfortunately, some models of remote afterloaders suffer from problems with source uncoupling and difficulties in source movement.

The sources in some of the units produce dose rates beyond the conventional low dose rate experience, and require adjustment for relative biological effectiveness.

CHAPTER 8

QUALITY MANAGEMENT FOR BRACHYTHERAPY APPLIANCES

All too often, in the press of performing brachytherapy procedures and the quality control involved with them, the applicators used in the procedure become taken for granted. Yet, the applicators play as important a role in the treatment as the sources or the treatment plan. Most of the quality assurance for brachytherapy applicators takes relatively little time while potentially preventing serious injury to the patient. Unless noted otherwise, the checks for applicator integrity should be performed before each use, usually at the time of packaging for sterilization.

For any applicators used with remote afterloaders where the channels contain built-in keys allowing only connection to particular parts of the applicator, acceptance testing of the applicator includes verifying that the keys work as intended. Currently, only intracavitary gynaecological appliances key to specific channels of the treatment devices.

Commissioning of any treatment appliance, be it a sophisticated MRI-compatible cervical applicator or a simple endobronchial tube, includes training appropriate personnel on the proper handling of the equipment, with particular attention to sterilization procedures. Disregarding the manufacturer's instructions often leads to unexpected, and possibly unnoticed, changes in the appliance, increased hazard to the patient and damage to the appliance, sources or other related devices.

8.1. INTRACAVITARY APPLICATORS

The most involved applicators find use in intracavitary gynaecological insertions. These appliances may serve many functions: positioning the sources or source with respect to the anatomy; shaping the anatomy; determining the penetration of the dose; adding space between normal tissue and the sources and shielding some of the normal structures. Incorrect performance of any of these functions results in erroneous doses and possibly serious complications.

8.1.1. Gynaecological appliances

The market place supplies many varieties of gynaecological appliances for treatment of cancer of the uterine cervix. This section addresses only the most common. The principles discussed translate easily to others since the general designs fall into a few categories. Except where noted, this section considers only afterloading applicators.

8.1.1.1. Uterine tandems. All the appliances for treatments of cancer of the cervix consist, in part, of an intrauterine tube that holds low dose rate sources or guides remote afterloading sources. Because the sources fall in a line, the intrauterine tube is called a 'tandem'. Figure 8.1 schematically illustrates the target volume for the treatments. The sources in the tandem supply the radiation to treat the cephalad portion of the target volume. The volume near the cephalad part of the vaginal fornices and the cervical os usually receive radiation from sources in the vaginal fornices if the anatomy permits placement in those locations. The main difference between the appliances arises from the vaginal source placement.

The insertion of the tandem carries with it the danger of perforation of the fundus (the cephalad wall of the uterus). Such perforation opens a route to infection in the perineal cavity, although few instances actually result. The perforation does cause the patient pain and some additional recovery time. While high dose rate treatments can proceed to completion following a perforation, such an accident is grounds for aborting the treatment for LDR applications, even after correcting the tandem, because the appliance remains in the patient for more than 2 days, increasing the probability of infection. To prevent a perforation, before inserting the tandem, the physician sounds the depth of the uterine cavity by inserting a narrow but blunt probe through the uterine canal. Placing a finger along the probe at the position of the external cervical os and holding it in place during removal allows the physician to transfer this depth to the tandem. Placement of a flange at this location on the tandem stops the insertion short of penetration into the wall of the fundus. The flange stays in place on the tandem by a small set screw (usually an Allen type). Two problems develop with the set screws:

- *The screw or hole strips.* With repeated use, the threads wear out either on the screw or hole. Nylon flanges with steel set screws particularly show wear. A crack in the flange through the hole presents an identical problem. Simply trying the screw in the hole at the time of packaging for sterilization detects this problem. This procedure also prevents packaging the wrong size screw for the hole.
- *The screw fails to engage the tandem.* Following chronic over-tightening, screws that directly tighten onto the tandem can wear flat spots on the tandem. The flat spot can prevent the screw from fully engaging the tandem. This results in a loose flange that fails to prevent insertion beyond the desired distance.

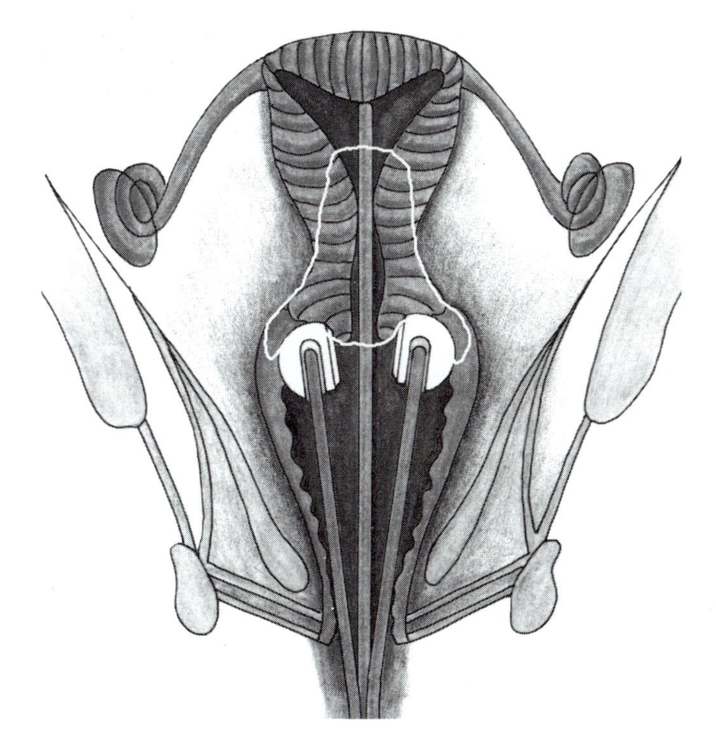

Figure 8.1. *A cutaway illustration of a uterus and vagina containing a tandem and ovoid. The white line indicates the target volume in treatment of cancer of the cervix. (Original figure without the line courtesy of Nucletron BV, Columbia, MD, USA and Veenendaal, The Netherlands.)*

Some flanges come with a keel or flag attached that runs parallel to the tandem. The keel helps maintain the orientation of the tandem. The weld that holds the keel to the flange requires inspection before packaging for sterilization to assure that the attachment remains solid. Likewise, tandems with rounded button closures on the deep end require inspection of the condition of the welds prior to packaging. Soft, non-metallic tandems pose much less danger of perforation. Instead of relying on an adjustable flange to stop the tandem, they have flanges built into the base of tandems of lengths selected to match that sounded in the uterine canal.

Tandems come with different degrees of bending, and the tandems for different types of appliance approach the bending differently. For example, the bending for a Fletcher–Suit applicator curls the tip around a given radius of curvature, and the greater degree tandems continue the curving farther along the tandem (see figure 8.2). The tandems for the Nucletron microSelectron, on the other hand, bend around a common hinge point forming two straight portions separated by an

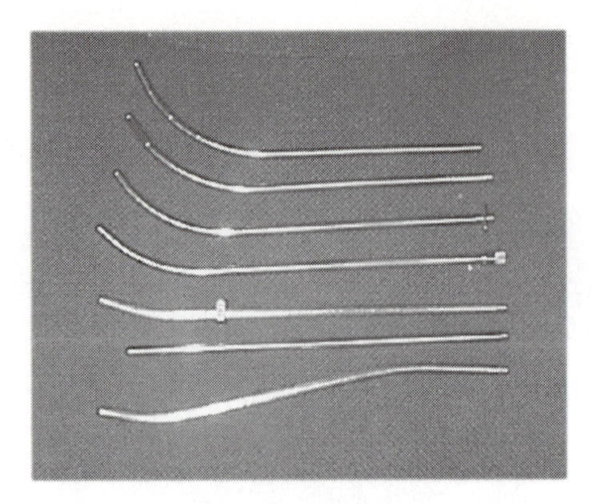

Figure 8.2. *A variety of tandems available for treatment of uterine disease. In practice, only a few of these actually find use in most clinics.*

angle. While the exact amount of curvature matters little in the treatment, changes in the angle affect the dose distribution, particularly the doses to the bladder and the rectum. For treatments based on either a dose atlas or previous calculations performed for the patient, the curvature must match that for the calculations within millimetres. The angle for an individual tandem can be changed unintentionally during insertion by running into a hard tumour in the canal, or during the therapy by patient movement. Comparison of the tandem with a life size silhouette template at the time of packaging checks for any deviations in the shape of the tandem, permitting correction before sterilization. Obviously, flexible tandems, such as with old Manchester tandems, obviate this problem, but also may produce very asymmetric, and unexpected, dose distributions.

LDR tandems close by a screw-on cap on the inferior end. This cap, as with all the other screws, should be tested for correct size and thread before packaging since those manufactured at different times or by different companies used various thread patterns. The threads also strip occasionally.

The sources in an LDR tandem fit into a plastic sleeve (commonly called the plastic tandem) that slides smoothly down the lumen. A plastic rod (called a pusher) fits down the sleeve to hold the sources in place. Both the sleeve and the rod come longer than necessary and require trimming: the sleeve to the length of the tandem (not all tandems are exactly the same length), and the rod to the length of the sleeve minus the combined lengths of the sources (see figure 8.3). The combination of the sleeve and rod should be tested for size by insertion into a tandem before cutting and loading. The tandem's cap holds the entire assembly together. If the lengths of all tandems in a facility are not the same, each tandem should bear identification related to its length for guidance in cutting the sleeve and rod. Cut too short, the sleeve becomes difficult to remove at the end of treatment; cut long the rod interferes with the closure of the cap on the end of the tandem,

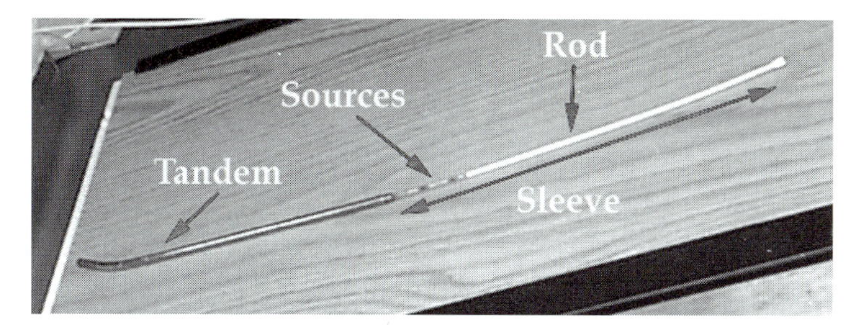

Figure 8.3. *Parts to loading a tandem. The sources are held in a clear plastic sleeve by a plastic rod. The rod and sleeve must be cut to size based on the number of sources used.*

and can result in the sources sliding out of their proper position, and even out of the tandem.

8.1.1.2. Fletcher–Suit family of appliances. Some of the most commonly used appliances follow the design based on the earlier preload-style Fletcher applicator. This appliance consists of the tandem and ovoids (as shown in figure 8.4). The ovoids fit together at a fulcrum, but usually have no fixed relationship with the tandem. By construction, the ovoids hold the sources approximately perpendicular to the tandem (i.e., with their axes roughly running in the anterioposterior direction). The ovoids have inner, hollow metal cylinders 1 cm in radius (2 cm diameter) with nylon caps that can increase this to 1.25 or 1.5 cm (or the diameter to 2.5 and 3 cm). During positioning, the physician uses the largest cap that fits into the vaginal cavity.

The ovoids contain tungsten shields to reduce the dose to the bladder and rectum. The different versions of the applicators have slightly different placement of the shields. Figure 8.5 illustrates the various orientations as taken from the landmark work by Haas *et al* (1985).

For loading, the sources fit into buckets on the end of handles that slide down tubes connected to the ovoids. Special hinging arrangements allow the source buckets to drop once they enter the interior of the ovoids, and then stand more or less upright.

The use of this appliance carries with it several caveats as noted in chapter 4. This present discussion considers only issues related to the integrity of the appliance proper.

- *Integrity of welds.* After repeated use and sterilization, the welds that hold the ovoid cylinders to the handles can become brittle and break. Were an ovoid to come free from the handle in the patient, sharp edges could cause lacerations, and the source could fall from the holder and lodge in the patient's fornices, or other spot, for an extended time, resulting in radiation damage. These welds

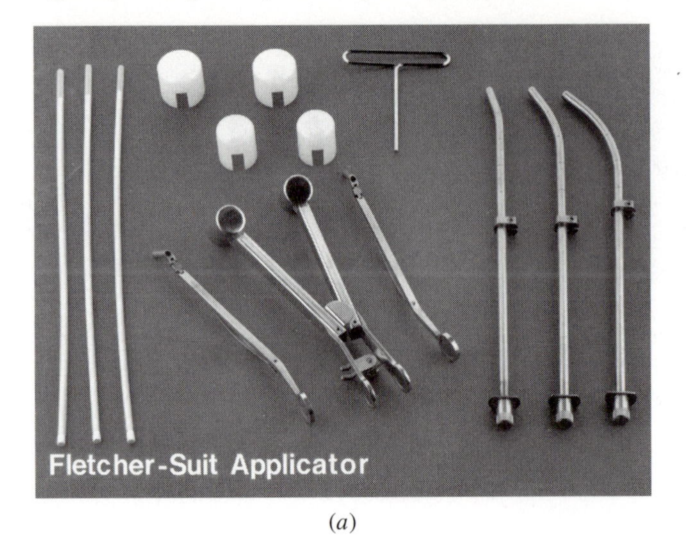

(*a*)

(*b*)

Figure 8.4. *The Fletcher–Suit applicator, consisting of a tandem and colpostats (ovoids). (a) shows the square handled ovoids that, while no longer available, remain in wide use (courtesy of 3M Corporation, St Paul, MN, USA). (b) shows the newer, round handled ovoids (courtesy of Best Medical International, Springfield, VA, USA).*

should be checked by pulling and trying to wiggle the ovoid on the handle. Any give should render the ovoids unfit until repaired (see the next item).

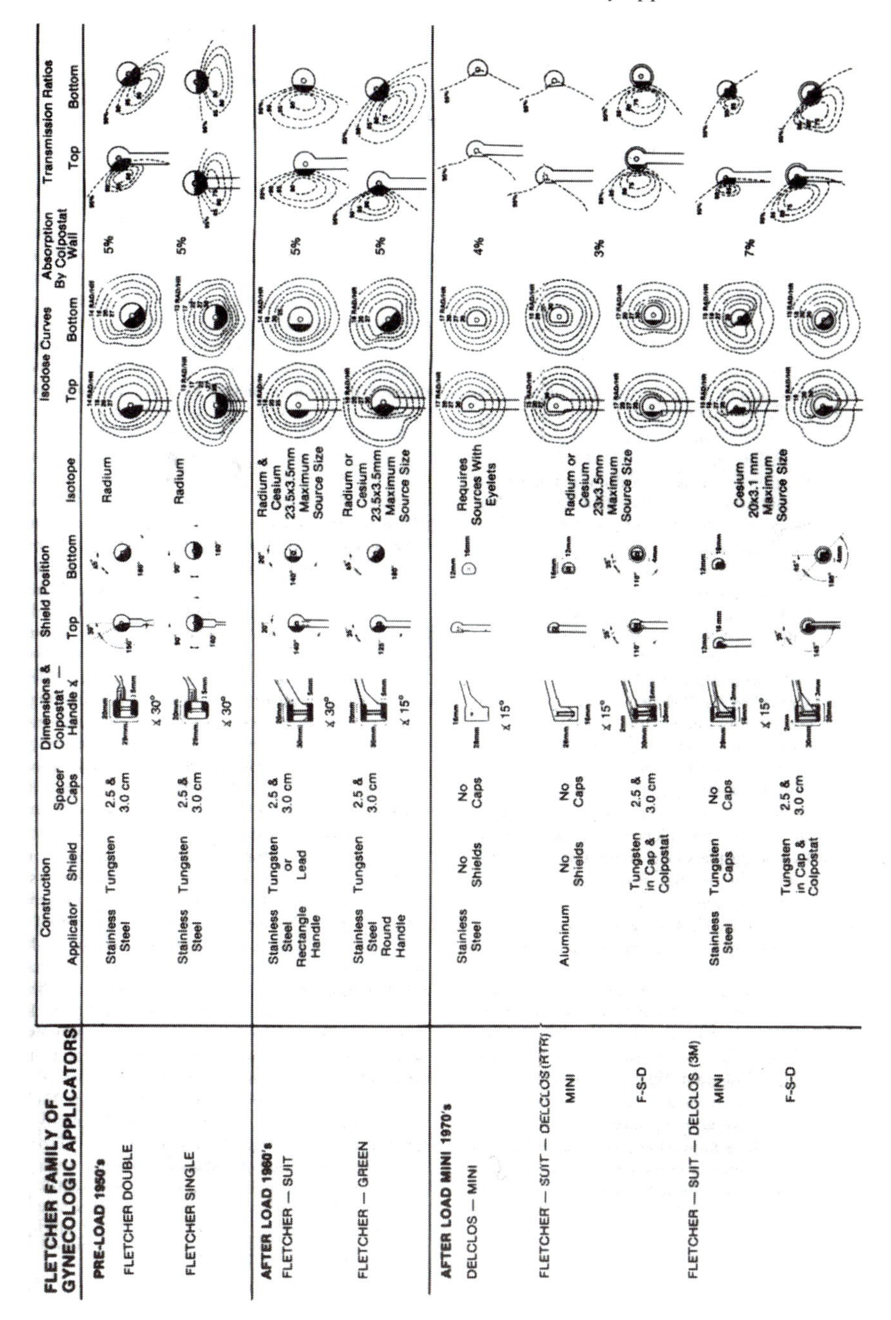

Figure 8.5. *Various formats for the ovoid shielding in, and the resultant dose distributions for, the Fletcher family of applicators (Haas et al 1985). (Figure used by permission of Elsevier, Amsterdam.)*

One manufacturer recommends immersing the ovoids in water and blowing compressed air into the handle looking for bubbles outside the ovoids.

- *Position of the shields.* Glue holds the shields in place. As with the welds, repeated sterilization can loosen the adhesive. Repairing the welds between the ovoids and the handles particularly stresses this glue. Loose shields that deviate from the proper location not only fail to reduce the dose to the sensitive normal structures, but can shield the target tissues. Checking the shield position is best performed with multiple radiographs. Unfortunately, a single radiograph often fails to provide the necessary information to evaluate the location of the shielding and several views may be necessary. The increased penetration of linac beams, compared to diagnostic quality beams sometimes aids in visualizing the shields through the rest of the ovoid material and can be useful during acceptance tests. Diagnostic energy radiographs taken at the same time provide references for periodic testing. Figure 8.6 shows radiographic images of Fletcher ovoids. This test should be conducted semiannually or after any repair on an ovoid.

The shields for the Fletcher–Suit–Delclos ovoids reside in the plastic caps instead of the metal cylinders because this appliance begins with a smaller naked diameter (often referred to as mini ovoids). Checking the location of these shields requires only visual inspection with a bright backlight. Since the plastic caps containing the shields should never experience high temperature sterilization or welding, the likelihood of drifting shield position becomes minimal.

- *Identification of the caps.* As discussed in chapter 3, the radius of the ovoids critically affects the dose distribution and the position of the tissues in that distribution. Because the caps cast little image on localization radiographs, markers in the caps coded to their diameter allow verification of the caps used for each patient during localization. Figure 3.1 shows a sample radiograph. The correctness of the markings should be checked each time they appear on a radiograph, since the markers can fall out of the caps.
- *Attenuation of the metal ovoids.* While the tandem seems to attenuate negligibly the radiation from either ^{137}Cs or ^{192}Ir sources, the thicker metallic ovoid wall produces a measurable effect. Measuring the attenuation of the wall proves challenging, but not beyond the resources available at most facilities. The determination compares measurements of the radiation exposure from a source first in free space, and then surrounded by the ovoid. Several aspects of this measurement require consideration. A suitably precise survey meter (ionization chamber based, not a Geiger–Mueller counter or scintillation detector) functions adequately for this measurement. The second reading is made with the source in the ovoid and the ovoid held in a clamp stand at the same distance from the meter. Care must be taken with the orientation of the ovoid to avoid the shielding shadow falling on or near the chamber. Sometimes the stability of the source or the ovoid held by clamps on a stand may not satisfy the demands of this experiment. In that

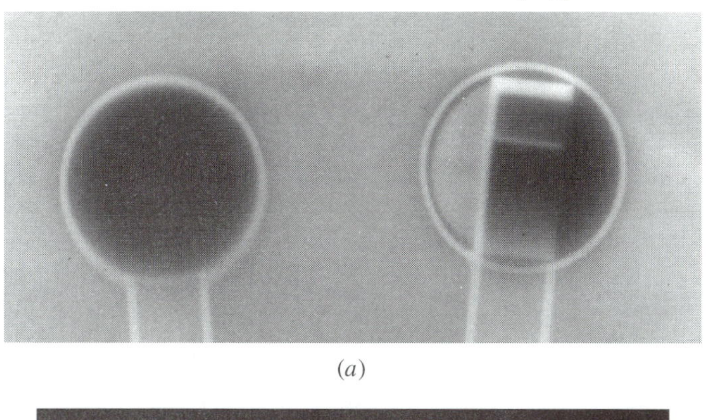

(*a*)

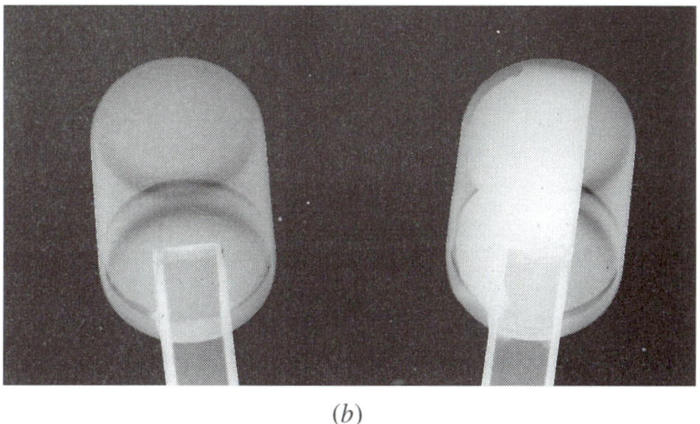

(*b*)

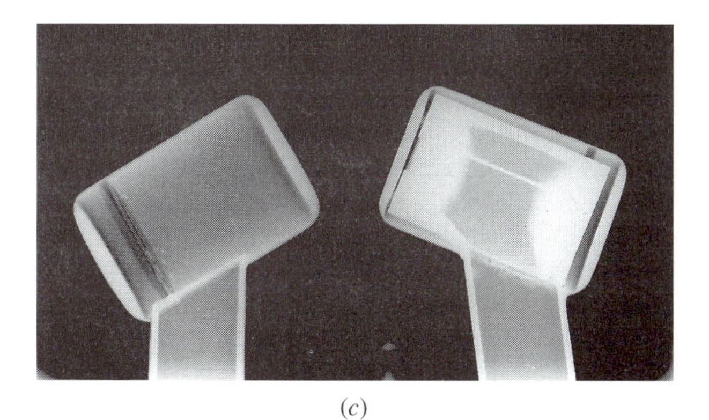

(*c*)

Figure 8.6. *Radiographs of Fletcher ovoids. (a) An image using a 4 MV linac beam collinear with the axes of the ovoids. The shielding from the ovoid on the left has fallen out of the applicator, and the bladder and rectal shields on the right cover the same angle. (b) A diagnostic radiograph taken at 120 kVp with the applicators resting on the film. While the image of the ovoid on the left still clearly shows the absence of shielding, the positioning of the shielding for the right ovoid becomes harder to assess. (c) The side view corresponding to the middle figure.*

case, imbedding them in the edge of a block of Styrofoam™ causes little perturbation on the outcome. The transmission of the ovoid cylinder simply becomes the ratio of the reading in the ovoid to that in free space. In calculations, the strength of the ovoid sources should be decreased by this transmission factor. For the Fletcher–Suit ovoids, Saylor and Dillard (1976) found a transmission of approximately 0.9 for the ovoids. This test needs to be performed only upon commissioning the appliance. The attenuation of the plastic caps differs little from tissue, and need not be evaluated.

- *Positioning jig accuracy*. The Fletcher–Suit–Delclos application connects the ovoids to the tandem with a 'bridge'. Any damage to the bridge can result in a skew of the ovoids. Again, this should be checked during packaging the appliance for sterilization[1].
- *Hinge function*. With repeated sterilization, the hinges on the ovoid source buckets become stiff, hindering efficient insertion of the sources. Occasional lubrication with a penetrating oil prevents this problem. (Unlike in radiation treatment units, the dose the oil receives remains low enough that polymerization does not become a problem.)

8.1.1.3. Other cervical applicators. *Henschke*. Next to the Fletcher family of applicators, the Henschke (see figure 8.7) is the most commonly used LDR appliance in the United States, and was the first afterloading gynaecological applicator developed. This applicator differs from the Fletcher in two important ways: the ovoid sources lie parallel to the tandem instead of perpendicular; and the inside of the ovoid is cut to allow placement in a smaller vaginal cavity while maintaining a larger spacer toward the lateral vaginal mucosa, and thus, better penetration in that direction (a feature incorporated into the naked metal cylinders of the Fletcher–Suit–Delclos applicator). All Henschke ovoids are plastic caps that slide over a stainless steel tube identical to the tandem. Unfortunately, the orientation of the ovoidal sources places the maximum intensity in the direction of the bladder and rectum, and the minimum toward the target. The sources load into the ovoids exactly as they do into the tandem, with the same precautions. The tandem and ovoids all connect at one point, which keeps the three components in a single plane in the absence of any damage to the bridge. The integrity of the bridge should be checked during packaging for sterilization, as well as all three screw-on closure caps. The regular Henschke contains no shielding in the ovoids to test, and the tube about which the ovoid caps slide attenuates the same as the tandem, and is evaluated in the same manner. Some modified Henschke applicators do contain ovoid shielding, but in plastic caps as with the Fletcher–Suit–Delclos applicator.

[1] It should be noted that for a 'good' insertion with a Fletcher–Suit applicator the tandem bisects the ovoid as seen on a lateral radiograph. By construction, the ovoidal sources for the Fletcher–Suit–Delclos applicator always hang posterior to the tandem on the lateral radiograph, potentially delivering a higher dose to the rectum.

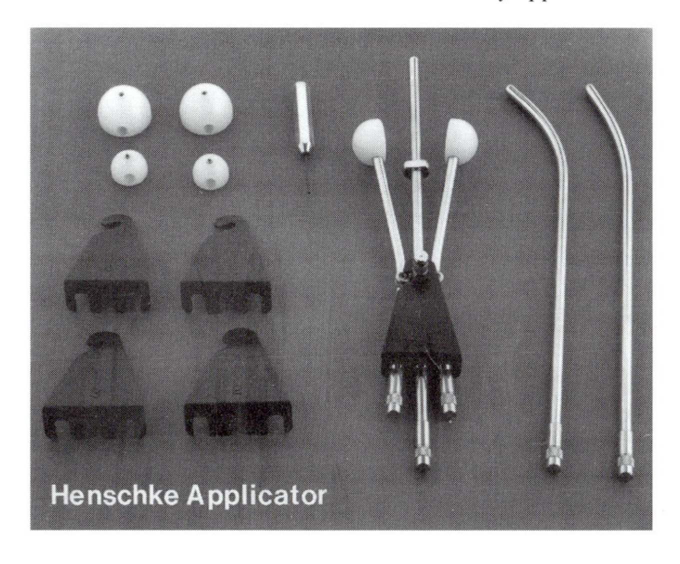

Figure 8.7. *The Henschke applicator. (Figure courtesy of 3M Corporation, St Paul, MN, USA.)*

Tandem and cylinders. When the vaginal cavity size prohibits the introduction of a tandem and ovoids, a physician often simply inserts a single tandem surrounded by concentric plastic cylinders around the tandem from the external cervical os to the labia minora (see figure 3.1). Such an applicator also serves to treat vaginal extension of cervical or endometrial disease beyond the reach of the dose from ovoids. The flange at the external cervical os anchors the cephalad end of the cylinders, and usually (but not always) a similar flange on the inferior side keeps the cylinders in place. As with ovoid cap size, the largest cylinder diameter the vagina accommodates results in the best dose distribution through the vaginal wall. Sometimes (but not always) the surface of the superior-most cylinder is rounded to fit more snugly into the top of the vagina. Other than imbedding markers into the outer surface of the cylinders to allow identification of the diameter on radiographs, and placing very thin metallic wafers between to visualize the starting and stopping points of each cylinder, the only quality control measures required for these applicators is to assure that the cylinders fit snugly over the tandem.

Tandem and ring. Used only with HDR applications, the ring serves the same function as the ovoids in the Fletcher or Henschke applicators, and generally has the source dwell at positions that simulate the location of the source in ovoids. The ring comes in several diameters to fit the size of the vagina. The tandem in the fixed assembly remains centred in the ring. The ring and tandem come in a variety of angles, as seen in figure 8.8. Plastic caps fit over the rings to provide spacing between the source track and the vaginal mucosa, increasing the penetration (and, thus, decreasing the vaginal surface dose) over simply using the

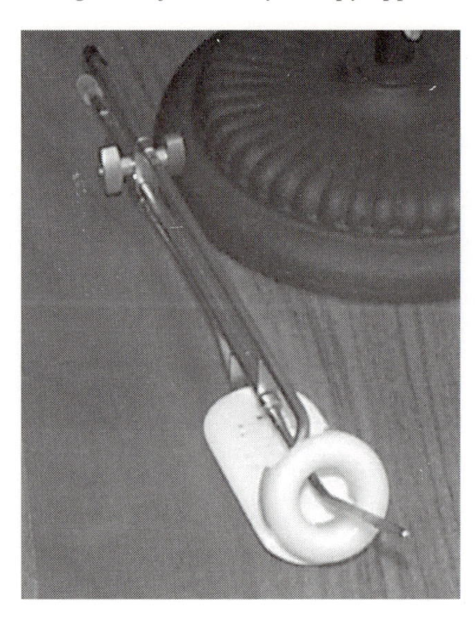

Figure 8.8. *The tandem and ring applicator, for use with high dose rate applications.*

naked ring tube. Unfortunately, most vaginas accommodate plastic caps of only about half the lateral thickness of an ovoid radius, resulting in less penetration (and a higher vaginal dose) than ovoids produce.

Unlike ovoids, the ring provides its own size indicator on radiographs. Since each size ring matches to one plastic cap, determining the ring size uniquely establishes the appliance geometry. The real challenge with the use of the ring applicator comes during identification of the true locations of dwell positions. As noted in chapter 6, the location where the source stops in curved tubes differs from the location indicated by the radiographic markers along a fully inserted cable. The difference arises because the source has support only at one end while the marker cable bridges between points where the cable touches the interior walls of the curve. Obtaining the correct locations for the dwell positions requires particular care.

Because the appliance fits together, the parts should be tested during packaging for sterilization. Bent tandems or rings require repair before packaging for patients.

The tandems come in fixed lengths (as measured from the cephalad surface of the ring) and that length should be marked on the package. While the physician should measure the tandem prior to insertion, writing the length on the package provides a redundant verification of the tandem length.

Noyes *et al* (1995b) discuss comparisons of optimized dose distributions for the tandem and ring compared with the tandem and ovoids, an article anyone considering changing from the tandem and ovoids to a tandem and ring should read.

The manufacture of the ring applicator suggests that the appliance should *not* be used for more than two years. Each instrument of this type should be labelled with an expiration date and this date should be checked before each use.

8.1.1.4. Vaginal applicators. Post-hysterectomy, prophylactic irradiation of the vaginal cuff forms one of the most common gynaecological brachytherapy procedures. The most common approaches to this procedure use either a set of ovoids without a tandem (since there no longer exists a uterine canal in which to place it) or vaginal cylinders, likewise without a uterine tandem. (See Noyes *et al* 1995a, for a discussion comparing treatments with cylinders and ovoids.) Similarly, a ring applicator can be used, also without the tandem. For vaginal cancer proper, or for cases where longer lengths of the vagina are at risk than just the surgical anastomosis, a cylinder must be used. The quality control measures for the ovoids or the ring follow the discussion in the section above.

The situation with the vaginal cylinders differs somewhat from that with a tandem and cylinders. With a tandem in the uterus, the geometry of the application may vary per patient, and often, per treatment. These variations in the configuration of the appliance, as well as differences in the curvature of the tandem, do not occur with the simple, straight-line treatments of the vaginal cylinders.

Treatment using any of these vaginal applicators, as with many applications for cervical cancer, defines the target by reference to the applicator and assumes that the appliance matches the patient's anatomy. Unlike the cervical appliances (other than the tandem and ring), when used properly, all of the vaginal applicators form a constant geometry, unaffected by variations in patient geometry. Thus, standard sets of treatment data (such as source loadings and treatment integrated reference air kerma—IRAK—for LDR applications or dwell weights for HDR) satisfy the treatment objectives for most patients.

Afterloading vaginal cylinders most often comprise a stubby, uncurved tandem with plastic cylinders that slide over the tandem (as in the case shown in figure 3.1). A snug fit between the tandem and the cylinders keeps the source track centred. If the hole becomes worn and the fit loose, the position of the sources becomes uncertain as does the dose to the patient. The fit should be checked during packaging before sterilization.

A rear flange holds the cylinders tightly abutting a flattened flange in the superior end of the tandem. The condition of the set screw for the rear flange should also be examined during packaging.

As discussed previously, the simple, in-line, tandem-based vaginal cylinder produces a dose distribution less suited to treat a vaginal cuff than ovoids, with the source axes perpendicular to the body axis and the two, side by side sources giving a broader dose distribution. For HDR applications, a vaginal ring serves as an alternative to ovoids. For LDR applications, the Wang applicator (figure 8.9) combines the better aspects of all applicators. This device uses a bucket-like holder for a cephalad source that allows the source to assume a position perpendicular to the cylinder axis. This source configuration was common with preloaded cylinders.

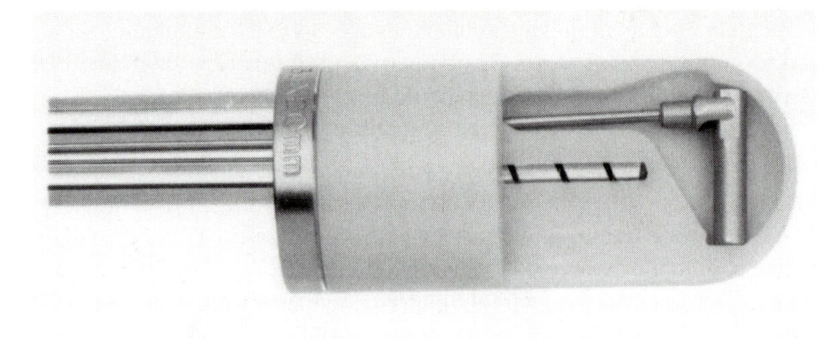

Figure 8.9. *The Wang vaginal applicator. Notice that the cephalad-most source resides perpendicular to the tandem axis, producing a dose distribution more conformal to the target with less dose to the bladder and rectum than with all the sources falling along the axis. (Figure courtesy of Mick RadioNuclear Instruments, Bronx, NY, USA.)*

As with Fletcher ovoids, the condition of the hinges for this cephalad source require monitoring, particularly following repeated sterilization.

8.2. INTRALUMINAL APPLICATORS

Most intraluminal applicators, such as for endobronchial or hepatic insertions, comprise a simple catheter. Those for endoesophageal or nasal treatments become more complicated, but usually have the simple catheter at the core of the applicator. For the simple catheters, integrity becomes the most important consideration. Were the end of the catheter to rupture, sources could leave the catheter and enter the patient. Smaller holes in the catheter allow body fluids into the source compartment. Such fluids can cause jamming of a high dose rate unit, or sliding of low dose rate iridium seeds along the nylon ribbon away from their intended positions.

Testing for small holes follows the procedure shown in figure 8.10. Placing the catheter mostly under water, increase the air pressure in the tube using a syringe, keeping an air-tight seal at the junction. Any bubbles forming on the exterior of the catheter warrant its rejection. To test the strength of the tip, apply a significant inward pressure on the guide cable just before insertion to ensure that the tip is solidly closed.

Endoesphageal and nasal insertions frequently place the source-carrying catheter inside a larger tube (e.g., an N-G or nasopharengeal tube, respectively). The outer tube provides both centring in the host cavity and spacing between the source and the mucosa to reduce the surface dose compared with the dose to the most distal target tissue. The main problem associated with such applications is failure of the inner tube to seat in the proper location in the outer tube, often due

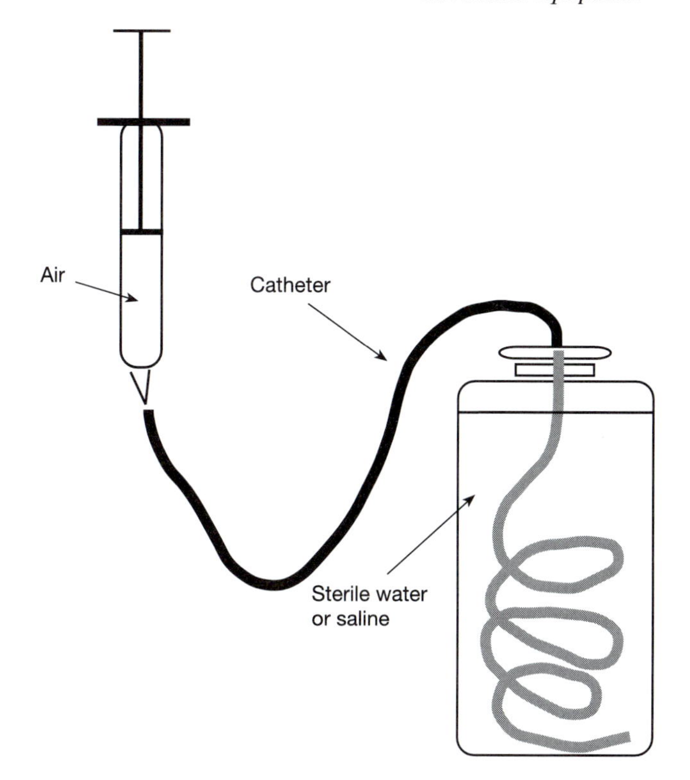

Figure 8.10. *A procedure to test intraluminal catheters for integrity before use.*

to binding from a fit between the tubes that is too tight. Before insertion into the patient, the composite applicator should be tested for easy passage of the inner into the outer tube.

8.3. INTERSTITIAL EQUIPMENT

While the equipment used for interstitial implants rivals intracavitary applicators for variety, most falls into the categories of needles, catheters or templates, and their associated accessories. The discussion that follows owes much to Sankara Ramaswamy, of Best Medical International.

8.3.1. Source-holding needles

This discussion only considers afterloading needles that hold source trains directly. Needles for insertion of catheters that are removed before treatment will be considered in the next section. Preloaded needles fall into the classification of 'sources', considered in chapter 2. Needles present the following critical characteristics:

(1) *Straightness*. Obviously, correct execution of a planned implant requires that the needle tracks follow the paths in the plan as closely as possible. Most implant paths follow straight lines. Implanting along straight lines proves difficult under the best of conditions with the assistance of a template. Needles often bend during insertion, sometimes through interactions with hard knots of tumour. Needles that start with slight bends produce a force vector in the direction perpendicular to the direction of insertion, increasing the probability of deviation from a straight path. Bent needles should be discarded at the end of a procedure; if they accidentally become packaged with good needles, the physician performing the implant should reject bent needles at the time of insertion. Once bent, a restraightened needle still has a higher tendency to bend again at the same point than a needle never bent.

(2) *Patency*. Sometimes a treatment plan requires a blank piece of ribbon inserted into a needle before the source ribbon, to act as a spacer and position the sources correctly in the target. Occasionally, after unloading the sources at the end of treatment, the needles undergo cleaning and autoclaving with these spacers left inside. The heat of the autoclave melts the ribbon that then adheres to the wall of the needle. If used in a patient, the glob of plastic in the needle prevents complete insertion of the source ribbon, and erroneous delivery of the dose. Checking patency before packaging the needles for sterilization simply entails insertion of a wire down the *entire* length of the needle. A mark on the wire corresponding to the correct depth of the needle lumen facilitates rapid assessment of possible blockages. While such blockages seldom pose a problem for the needles used with permanent implants (since the needles only find use once), roughness of the inner wall of permanent implant needles causes sources or spacers to bind and jam (preventing delivery of the sources), or hang (resulting in delivery of the source at an unintended location along the needle track during withdrawal). While no widely accepted procedure to screen needles for roughness exists, dropping a spacer down a needle just prior to loading sometimes identifies potentially problematic needles.

(3) *Integrity*. The ends of interstitial needles, just as with intraluminal catheters, must seal off body fluids from the source or sources (except for delivery of permanent implant sources). For steel needles this seems not to be a problem. However, the tips of flexible, plastic needles, designed for use with templates, not uncommonly break during insertion from the pressure of the stylet. On occasion, they have broken off at the skin (causing difficulty in removal), and not uncommonly at the template. At the time of writing, there are no quality assurance procedures reported that help eliminate these problems, but the following recommendations may provide the best assurances possible:

- *check for cracks or nicks,*
- *check the tip for sharpness and integrity,*
- *check the gap between the obturator and the tip against a light,*

- *check the luer lock for proper fixation and fit of closure cap and*
- *check each needle for proper fit and fixation in the temple holes.*

(4) *Sharpness.* The desired sharpness for an interstitial needle depends on the intended use. For needles that pass through the patient, entering on one side of a target and leaving on the other, such as breast implants, a very sharp tip cuts more cleanly through the tissues and often yields a straighter insertion[1]. A dull needle tends to deflect off regions with increased density. However, very sharp needles carry with them an increased probability of cutting through an artery. This becomes a serious consideration in neck implants. Sharp needles also cause problems for implants where the tips of the needles remain in the patient, such as gynaecological template cases. For these patients, sharp tips continue to lacerate the tissues in contact, causing great pain, particularly with movement. Even very heavy sedation and pain medication fail to relieve the suffering in these cases. The tips of such needles should be pointed enough to part tissues with some insertion pressure, but not so sharp as to cut a finger pressed, but not forced, onto the point. Each needle should be checked for the proper sharpness for its intended use prior to packaging for sterilization. The person packaging the needles should ensure segregation of needles by sharpness.

(5) *Bevel.* As with sharpness, the desired bevel for an implant needle depends on its application. In general, needles with tips remaining in the patient over the duration of the treatment use blunted, conical points, while those that exit the patient may use bevelled tips. (Degree of bevel is discussed below for catheter insertion tools.) Quality control for needle bevel simply entails ensuring that all needles packaged together share the same bevel angle.

(6) *Diameter.* The diameter of an implant needle becomes important in two aspects of the treatment: the *inner diameter* must allow passage of the source yet fit it snugly to minimize positional, and therefore dosimetric, uncertainty; and the *outer diameter* must fit snugly through any template or other positioning device. For the most part, inner and outer diameters correlate, with needle walls about 0.23 mm thick. Both these aspects of the needles require acceptance testing upon receipt, and checking during packaging to prevent mixing of needle diameters in a single package. The check of the outer diameter can use a gauge, while the check for patency can also check for the inner diameter by using a wire that just fits the inner diameter.

(7) *Length.* While critical for high dose rate implants, accurate needle lengths form an important component for all implants. As with needle diameter, both the inner (somewhat corresponding to the active length in older, preloaded needles) and outer length (the physical length) of a needle affects implant ac-

[1] After completing the implant and setting the needles in a template, the sharp ends need a cover to prevent injury to the patient from rubbing against the tips.

curacy, and particularly the relation between the two. Positioning the needle in the patient often relies on setting the position of the needle tip under some form of guidance (such as fluoroscopy, ultrasound or palpation) or inserting the needle with a predetermined amount of the end out of the patient. Either method assumes a constant relationship between the tip of the needle (or the end) and the location of the source when inserted into the needle. While imaging with markers in the needles demonstrates any deviations from this assumption (based on accurate simulation of the source by the markers—see chapter 9), large implants often render simultaneous imaging of dummies in all needles difficult to impossible. The relationship between the needle tip and the position of the source for LDR insertions and end-seeking HDR units depends on the length of the inactive end comprising the needle tip plus any nonactive material in the source train. HDR units that specify the length to the first dwell position require constant physical lengths to position the source the same in all needles using the same value for the dwell locations. Again, the lengths of needles must be checked as part of acceptance testing, and verified to prevent mixing lengths at the time of packaging for sterilization. The wire-insertion check for patency can simultaneously check for length by marking the wire with a coloured band at the correct inner length.

(8) *Connector coding (for HDR).* High dose rate needles attach to the treatment unit through an intermediary transfer tube that guides the source from the unit to the needle. The needle must lock into the transfer tube, or the unit will not advance the source past the junction. Failure to make this connection could result in incomplete treatment of the target. That the connectors function properly forms an important part of the acceptance testing for the needles, and should be performed at the same time as the length determination.

(9) *Collars (for templates).* Template needles (except for guide needles) have additional collars around the ends large enough in diameter not to fit through the template holes. At the end of treatment, following unloading of the sources, as the physician pulls the template from the patient, the template pulls all the needles out of the patient with it. These collars should be checked for integrity during packaging for sterilization.

8.3.2. Catheters

Implants based on flexible catheters to carry the source or source ribbons make considerable use of peripheral equipment, all of which must function well or compromise the quality of the procedure.

8.3.2.1. Insertion needles. The insertion needles cut the track through the tissue that the catheter will follow. Some of the desirable characteristics of these needles follow those for the implant needles in the previous section (for the same reasons), while some differ markedly.

(1) *Straightness.* As with the source-carrying implant needles, bent catheter insertion needles reduce the conformality of the implant, and should be rejected (except for curved needles used to enter and leave the patient through a flat surface, of course).

(2) *Sharpness.* Catheter-insertion needles do not remain in an unanaesthetized patient, so extreme sharpness poses no comfort problem. Extremely sharp needles often cut tracks with less trauma to the tissue, so can result in less postoperative discomfort. Sharp needles also tend to follow the intended direction better than dull needles, particularly through dense tissues. On the other hand, near arteries, sharp needles present a high likelihood of cutting the artery, while dull needles often push arteries aside. The sharpness of the needles should match the characteristics of the anatomy. Sorting by sharpness before packaging for sterilization prevents unintentional use of the inappropriate needle sharpness.

Because most needles used as guides for catheter-based implants are open ended in the front, they sometimes tend to 'core' the tissue they pass through, rather than separate it. This inhibits that passage of the needle, and forms an effectively duller cutting edge than a solid tip (not to mention causing considerably more tissue damage). Insertion of a sharp stylet through the needle such that the tip of the stylet aligns with the cutting edge of the needle counteracts this effect. The stylet also stiffens the needle by making it more solid, and reduces the likelihood of the needle bending and deviating from its intended path. The stylet can also provide a conical cutting edge (see below under bevel) while the needles alone cannot. Other than assuring its sharpness, the stylet needs no other quality control.

(3) *Bevel.* Larger degrees of bevel for a needle tip translate into sharper cutting edges. The discussion above considered the occasions for choosing various sharpnesses for the needles. However, the bevel of the needle tip results in a component of force perpendicular to the direction of insertion (see figure 8.11) that pushes the needle away from the intended track. The cross-needle force increases with bevel angle, but with large angles, the increased cutting power of the tip more than compensates for this force. That conical tips do not generate such a force is one of the advantages of using a stylet. While, as with source-carrying implant needles, quality control for needle bevel simply entails ensuring that all needles packaged together share the same bevel angle, extremely bevelled needles form a hazard for patients and staff. As with scalpel blades, a slight brushing of the needle tip against flesh can produce a serious laceration.

(4) *Diameter.* Outer diameters of catheter guide needles should approximate the diameter of the catheter, but the exact relationship bears little on the ease or quality of the implant. The outer diameter only becomes critical

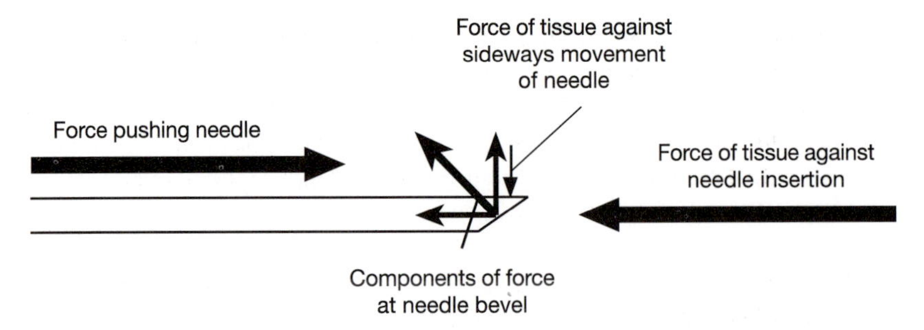

Figure 8.11. *Forces on a needle tip passing through tissue. The bevel translates the force of the tissue opposing the passage of the needle into components along the needle (resistance) and perpendicular to the needle (deviation). The tissue also produces a force counter to the sideways movement of the needle.*

when a template guides the needle positioning. In this case, as with the implant needles, the fit must prevent significant deviation of the needles during insertion. (See the entry on templates.) The inner diameter need only accommodate the part of the catheter that will pass through it. The passage of the catheters though the needles needs testing before packaging of supplies for the operating room to ensure availability of the proper equipment during the procedure.

8.3.2.2. Catheters. While the catheters proper form one of the most important parts of the catheter-based implant, few reports of failures or problems exist. However, some aspects bear review before the procedure.

(1) *Integrity*. If the lead end of the catheter will remain in tissue or in a body cavity (such as commonly the case with mouth implants), the integrity of the fluid-tight seal on the tip becomes critical. The test for integrity can be performed sterilely as for an endobronchial application (as in figure 8.10) or before packaging for sterilization. Implant catheters, unlike endobronchial catheters, experience little force on the tip during insertion. Slits in the side of the catheter also pass fluids and produce the same problems as failures of the tip, but almost always show visibly.
(2) *Diameter*. The inner diameter of the implant catheters must satisfy the same criterion as source-carrying implant needles: that the source fit the lumen snugly. This should be tested with a dummy wire before packaging for sterilization for the procedure. The outer diameter only becomes a factor if the catheter body slides down the guide needle, as for 'blind end' implants, where the tip of the catheter remains in the patient and the needle slides out over the implanted catheter. For most implants, only the narrow tail of the catheter need fit through the needle. Any question about the clearance

for either operation should be resolved well before the procedure to allow ordering of compatible equipment.

(3) *Length*. The length of the catheter need only cover the implant. A simple measurement during planning assures this. However, for loop-type implants, the catheter covers the target in three directions (in, over and back), and the total must include all this travel.

8.3.2.3. Buttons. Buttons hold the catheters in place in the patient and often fix the source ribbon in place in the catheter. Buttons come in two varieties: plastic and metal.

- *Plastic buttons*. Plastic buttons serve first to hold the catheter immobile between implantation and loading. The buttons must fit very snugly around the catheter to prevent the catheter from slipping in the patient, yet must slide to allow repositioning of the catheter if necessary. They must hold the catheter without significantly narrowing so the source or source ribbon may pass into the lumen. For high dose rate treatments, only these buttons may hold the catheters in place, in which case a drop of 'super glue' between the button and the catheter adds to the fixation. As a second function, the plastic buttons help identify groups of catheters through colour coding. Because of the critical nature of the fit between the buttons and the catheters, these pieces should be tested for compatibility during the packaging for sterilization.
- *Metal buttons*. As with the plastic buttons, the metal buttons may serve two purposes. As the first function, the metal buttons provide the final closure and fixation for the catheters. After loading the source ribbons for LDR implants, crimping the button onto the catheter and the ribbon together holds the sources in the proper location. The crimped buttons also hold the catheters more firmly than the plastic buttons. The crimp, if performed with Kelly clamps, may be reversed in most cases by compression orthogonal to the original crimp to allow correction in the ribbon position if necessary. HDR implants require the source to pass into and out of the catheter, and, thus, crimping, even of a light, temporary nature poses a potential for preventing the execution of the treatment. The second function of metallic buttons is to shield partially some tissues in line with the catheters. An example of this use includes shielding the tongue from some of the radiation from a floor-of-mouth implant. When used for shielding, gold or tungsten buttons, rather than the normal stainless steel, provide more attenuation due to their higher atomic number and higher density. Metal buttons need only slide over the catheters, and other than checking for this during packaging, require no other quality control measures.

Most catheter-based implants utilize catheters with buttons previously attached to the tip (and in these cases, blind) end. The buttons on these catheters should be tested for fixation by pulling with considerable force, but not enough to stretch the catheter, to assure that the button will not move once implanted in the patient.

This test should be performed *just* prior to implantation to include checking for any degradation due to the sterilization process.

8.3.3. *Flexible needles and fixed-length plastic tubes*

Flexible implant needles form a class of applicators between ordinary, metal needles and plastic catheters. These needles usually serve in template applications. In addition to the problems discussed above in the section on needles, the plastic in these needles, and in several plastic tube-type applicators (such as Heyman capsules or Simon–Silverstein applicators) can shrink during sterilization (Slessinger 1995b). This becomes a problem for treatments with remote afterloaders that require a specified distance from some point in the unit to the tip of the needle or applicator in order to correctly position the source. Just prior to insertion into the patient, the inner length of the tube should be measured with a calibrated stylet. Those needles or tubes falling outside the tolerances for positioning the source correctly can be discarded before insertion.

8.3.4. *Templates*

Templates serve to assist placing needles in the proper location in an implant, and often holding the needles in the desired location. Guiding the needles to their planned destination requires three characteristics of the template holes:

(1) *Proper hole placement*. The hole pattern must match that used by the computer in generating the treatment plan. Deviations from the treatment plan most often result from using the incorrect magnification for the hole pattern (or not measuring the separation between holes and making assumptions concerning the dimensions), rotating in a circular pattern differently during implantation than during planning or using the wrong template altogether. The correspondence between the patterns in the computer and the template identification must be checked when creating the computer file, if the file is stored, or each time the pattern is entered. The orientation needs verification just prior to insertion of the needles.

(2) *Proper hole angulation*. Some templates, particularly those for prostate implants, contain holes angled to avoid bony or sensitive structures. Small errors in the angles of these holes produce large deviations in the position of the needle tip 15 cm or so deep. As a visual check of the needle positions, place needles through all the holes and observe the needle pattern at various distances from the template, and compare with the desired pattern. Even for templates with all the needles parallel, an initial check during acceptance testing inserting all the needles verifies that the holes produce the planned result.

(3) *Sufficient guidance for the needles*. To position the needle tip at the desired location not only requires that the needle starts at the proper location at the template, but that the template can guide the needle tip accurately. In

addition to the angle of the hole checked in the previous item, the diameter and length of the holes combine to determine the allowed deviation of a needle passing through the hole. The equation describing the acceptance angle follows that for the acceptance angle of a radiographic grid. Looking at figure 8.12, the acceptance angle for the needle becomes

$$\theta = \cos^{-1}\left\{\frac{d/D + (L/D)\sqrt{1 + (L/D)^2 + (d/D)^2}}{[1 + (L/D)^2]}\right\}.$$

This angle decreases as L increases or as d approaches D. The guidance need not come from one long hole. Two holes separated by a distance equivalent to L in figure 8.12 serve the same function, as shown in figure 8.13. Deviations caused by tissue irregularities usually exceed any uncertainty due to slope in the acceptance angle. However, poor guidance caused by lack of constraint by the template hole adds another, unnecessary variable leading towards low quality or unsatisfactory implants. The match between the needle diameter and the hole size should allow the needle to pass only with some force exerted. Loose holed templates should be rejected during acceptance testing, and retested during packaging for sterilization before each case.

In a sandwich template, as shown in figure 8.13, only the entry-plane plate of the template provides guidance to the needle. The deep plate helps the person inserting the needles know whether the needle exits the patient at the correct location. Sometimes, if a needle persists in deviating from the exit hole, changing the bevel of the needle (or from a conical to a bevel, or vice versa), or rotating the direction of the bevel changes the path of the needle. Since the distal plate provides for the alignment of the needles, the holes in the proximal plate should be slightly larger than normal, allowing for some modification in the aim of the needles so to hit the target hole exiting the patient. The greatest difficulty in the use of sandwich templates (after hitting the opposite hole) is aligning the two plates. The brace between serves this function, and requires considerable care in manufacture. Before packaging the template for sterilization, the entire template should be filled with needles and the resultant pattern compared to the treatment plan.

(4) *Sufficient fixation for the needles.* When all needles in an implant occupy their correct location, the template then must hold the needles in place and be resistant to normal forces pushing or pulling them. Templates accomplish the fixation mostly in one of two ways. Many templates, such as the gynaecological Syed–Neblett, encircle the needle holes with O-rings between two plastic plates. After placing the needles in such templates, tightening screws connecting the two plates squeeze the O-rings, and they, in turn, squeeze the needles, preventing their movement (Feder *et al* 1978, Fleming *et al* 1980). For custom designed templates, sandwiching a layer of Superflab™ between two acrylic sheets connected by screws serves the same function (Ritter *et al* 1989). Some disposable templates consists entirely of soft plastic. These

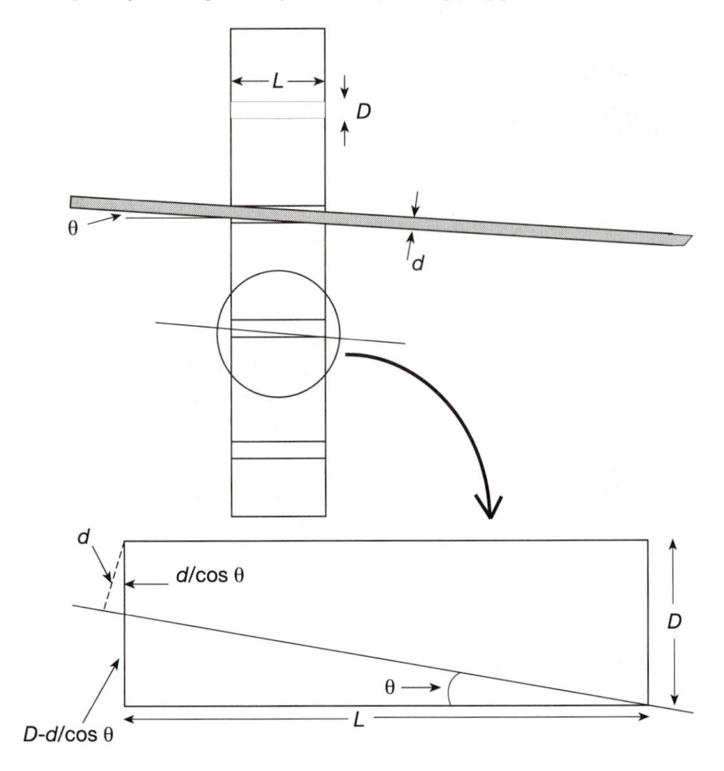

Figure 8.12. *A schematic diagram showing the acceptance angle for a needle passing through a template.*

templates continually grab the needle and prohibit movement. Needle insertion and removal utilize alcohol as a lubricant, allowing the needle passage through the template hole until the alcohol evaporates, at which time the needle remains fixed. The other common method to secure the needles in the template uses individual set screws for each needle. The Syed–Nebett urethral template uses this approach. That this method requires tightening (and reaching) a screw for each needle limits it suitability to implants with a small number of needles. In addition to these methods of needle fixation, for implants guided by templates on the entry and exit surfaces, collars on the ends of the needles stop inward travel at the entry template plate. Fixation in the outward direction can be accomplished by bending the needle tips at the distal template plate. Removing the needle from the patient requires either straightening the needle tip again or cutting the unloaded needle between the distal template plate and the patient, possibly leaving a rough edge that could injure the patient beyond what is necessary for the implant.

The fixation provided by the template seldom fails suddenly. Rather, the friction degrades slowly with time and use. After evaluating the template

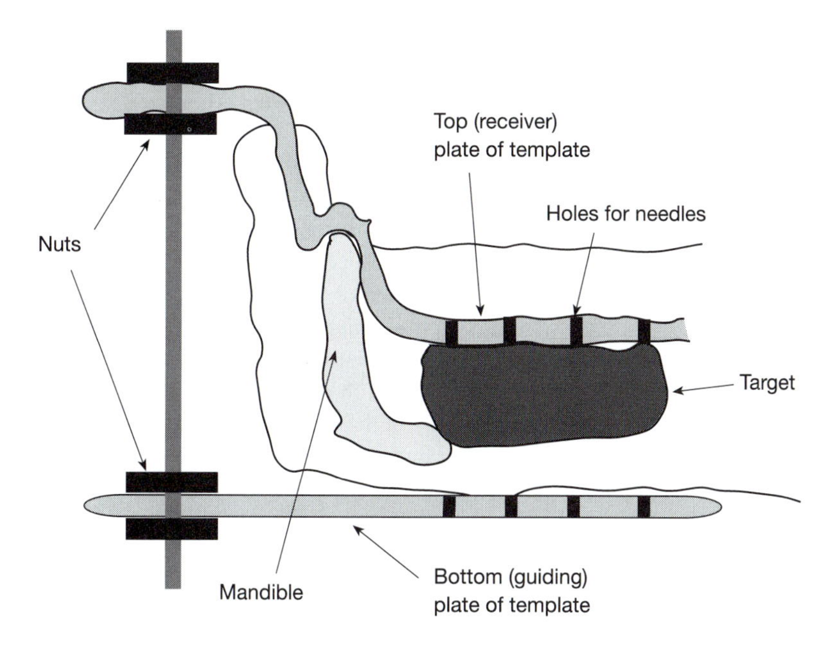

Figure 8.13. *An example of a sandwich template. The implant needles enter through the bottom plate, pass through the target and exit through the holes in the top plate.*

for fixation during acceptance, simply testing the fixation during each use provides the quality assurance for the next case. Replacement of 'O'-rings for templates using them should be a routine part of periodic maintenance. The screws that tighten two halves of a template also need to be checked for condition and operation during packaging. If the template uses rods to guide the two halves, the condition of the rods also should be checked for rust or bends.

Because HDR and LDR needles differ in diameter, using the one modality needle with the opposite modality template either results in poor fixation or failure to insert the needles through the template in the operating room. Thus, the person in charge of keeping the equipment needs to pay particular attention to separating and clearly labelling the needle sets.

Templates that use an obturator require the following additional checks:

(5) *Correct fit of the obturator in its hole.* Through damage, excessive wear or mismatch from different template sets, the obturator may not fit well (if at all) in the corresponding hole of the template. This should be checked at the time of packaging for sterilization.

(6) *Proper operation of the set screw holding the obturator.* As with the other set screws discussed, the set screw that holds the obturator in place in the

template may strip, crack the plastic that holds it or fail to hold the obturator tightly. Again, this should be checked at the time of packaging for sterilization, along with the match between the enclosed Allen wrench and the screw.

(7) *Proper fit of a tandem through the obturator*. If the application uses a tandem passing through the obturator, the fit and fixation become critical and require testing during packaging. Great care must be taken when using a tandem with a template. The tandem tends to curve anteriorly toward the cephalad portion of the volume, passing between or in front of the anterior implant needles. Because the two parts of the appliance, the tandem and the template needles, behave so differently, the dose distribution frequently contains very high and very low dose regions. In addition, the strength of the sources in the tandem tends to overwhelm that of the needles, again making a uniform distribution through the target volume unlikely. Selectively unloading the tandem earlier creates a biologically complicated situation. Often, a better approach simply treats the entire target volume using only template needles, including possibly one passing through the centre of the vaginal obturator.

Templates used with ultrasound-guided implants require alignment with the ultrasound (US) image. Initial misalignment proves the rule rather than the exception. Two tests check important correspondences:

- *Rotational alignment*. With the US probe inserted into a bucket of water, place a needle in a central hole until it produces an image on the monitor. Compare the location of the centre of the needle's image with that of the indicated grid point associated with that hole. Since the water causes no deviation of the needle, the images should correspond to within 1 mm. On some systems, rotating the US probe in the harness corrects the deviations. However, on most systems, the probe locks into place in the harness, and the rotation indicates an undesirable rotation of the transducer element and the need to replace the probe.
- *Scaling*. Assuming a correct rotational alignment, continue with the same setup, and insert needles into the four corner holes of the template and compare their images with the indicated locations on the grid. Again, the centres of the images should match the indicated grid positions to within 1 mm. A failure indicates erroneous scaling of the image. Some units permit adjustment of the image scaling either through software or hardware control.

8.4. SURFACE APPLICATORS

Surface applications vary so widely that covering all possibilities exceeds the scope of this text. The discussion below considers only two particular types of application.

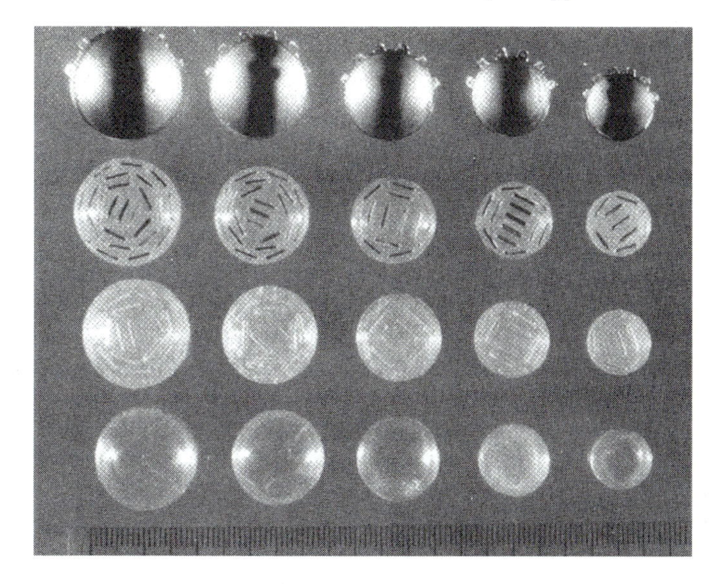

Figure 8.14. *Eye plaques developed as part of the Collaborative Occular Melanoma Study. (Figure courtesy of Sou-Tung Chiu-Tsao.)*

8.4.1. Eye applicators

Ophthalmic applicators used for treatment of pterygium consist of ^{90}Sr sources on a handle. These applicators actually are the source, and the discussion in chapter 2 covers them.

Applicators used for treatment of choroidal melanomas use seedlike sources in a plaque sutured on the surface of the eye. Figure 8.14 shows a set of such applicators used in the clinical trials of the Collaborative Ocular Melanoma Study. The different sizes match the size of the target. The applicator consists of three parts: the sources, the carrier and the backing. The carrier holds the seeds in the predetermined position, while the backing provides rigidity for the carrier and radiation shielding for persons attending the patient (and other parts of the patient as well).

The Silastic™ (silicon rubber) carrier contains impressions cast during manufacture that hold the sources. Because of the curved surface of the carrier, checking the location and orientation of each of the slots becomes difficult. Fortunately, the manufacturing process, while open to other failures, seldom (if ever) places the holes incorrectly, and slight errors in one or two slots produces minor errors in the dose distribution. Most problems with a carrier become obvious during loading of the sources into the slots. Examples of problems include incompletely cut slots that prevent seating of the source, or erroneously large slots that fail to position the sources firmly. Encountering a problem with the carrier calls for replacing the carrier with a different one, since they are disposable.

During loading, a light layer of silicon rubber cement holds the sources in place, yet loaded sources can work loose due to the pressure of loading subsequent sources. After loading, all sources should be rechecked for seating properly in the bottom of their respective slots.

The gold backing should begin clean and smooth on the inside, free from the cement of previous treatments. Any dents prevent proper seating of the carrier in the backing, or fit of the carrier against the patient's eye, and warrant replacement of the backing before the treatment. Cement in the suture holes hinders threading the suture through the holes after the suture lines have been anchored in the eye based on the dummy plaque hole positions. Before packaging the plaque for sterilization, a needle should be run through the eyelets to assure their patency.

Following sterilization, the plaque and the dummy plaque should be checked visually for integrity and possible damage in case inappropriate sterilization methods were used (i.e. high temperatures).

8.4.2. Skin applicators

An astounding variety of surface applicators for treatment of skin diseases appear in the literature. Some of these applicators form reusable and fixed devices with standard dosimetry, while others conform to an individual patient's treatment. In general, parameters to check common to all surface applicators for skin include:

- *Thickness.* The dosimetric calculation for a mould assumes a specific distance between the centres of the sources and the skin surface. The radius of the sources, and any surrounding material such as the ribbon around ^{192}Ir sources, add to the thickness of the mould proper that holds the sources to give the distance.
- *Source position.* While source position on the mould obviously influences the dose distribution, planning systems often allow placements for dosimetric calculations impossible to achieve in reality. For treatments with the sources sitting on top of a thick mould, small deviations from the plan make little difference in the delivered dose distribution. However, with thin moulds, the positioning becomes much more critical.
- *Source fixation.* Particularly with homemade, *ad hoc* moulds, keeping the sources in place may prove the most difficult part of construction of the applicator. Some good mould materials (e.g. wax or Superflab) stick poorly to tape and many adhesives. Sandwiching the sources between two custom formed plates held together by screws forms an ideal device for holding the sources fixed, but requires considerable time, material and facilities to fabricate. However the sources are held in place, before placement on the patient, the sources should be pushed at and prodded to ensure that they do not move or fall out of the mould.

For reusable skin applicators these tests occur as part of acceptance testing; for individual applicators as checks prior to placement on the patient.

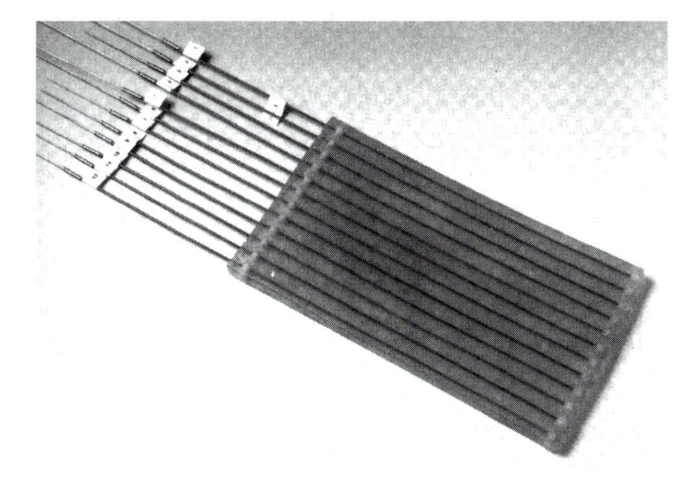

Figure 8.15. *An example of a HAM applicator for high dose rate intracavitary brachytherapy. (Figure courtesy of Mick RadioNuclear Instruments, Bronx, NY, USA.)*

8.4.3. Intraoperative applicators

Intraoperative brachytherapy procedures consist of placing catheters and treating during a surgical opening of the patient. The delivery of the therapy during the surgery differentiates intraoperative brachytherapy from operative placement of catheters, followed by closing the patient, for later treatment, often in several fractions. The latter follows the same guidelines as a catheter-based implant, while most intraoperative treatments present as special cases of surface applications. One of the most developed and studied intraoperative applicators is the HAM applicator, named for the originators, Louis Harrison and Lowell Anderson at Memorial Sloan-Kettering, and Felix Mick of Mick RadioNuclear Instruments. Figure 8.15 shows a sample HAM applicator, but they come in a wide range of sizes. The applicator uses catheters imbedded in a block of flexible plastic. Because the treatment takes place with the open patient, the delivery always uses an HDR source. The placement leaves little time for detailed dosimetry, so treatment dwell times follow from precalculated plans in an atlas based on the size of the treatment applicator and degree of curvature of the catheter plane. Other than verifying the size of the applicator and its curvature, and assuring the use of the proper atlas entry, the quality control procedures follow those for a catheter-based HDR implant.

8.5. SUMMARY

Table 8.1 summarizes the quality control checks suggested for various treatment appliances.

Table 8.1. *Quality control procedures for brachytherapy appliances.*

Article	Feature to be tested	Frequency
Gynaecological appliances		
Tandems	Flange screws function	Each use
	Curvature	Each use
	Closure caps function	Each use
	Plastic sleeve and rod fit and slide	Each use
Fletcher-type ovoids	Source carriers function	Each use
	Integrity of welds	Each use
	Position of shields	Semiannually; after repair
	Identification markers	Each use
	Attenuation of ovoids	Acceptance
	Bridge integrity/thumb screws	Each use
Henschke applicators	Placement of caps on stem	Each use
	Bridge integrity, screws in bridge	Each use
	Closure caps function	Each use
Tandem-based cylinders (and tandem checks)	Flanges function	Each use
	Identification markers	Each use
	Cylinders fit snugly	Each use
Solid cylinders	Source carriers function	Each use
	Closure caps function	Each use
Intraluminal catheters	Integrity	Each use (after sterilization)
	Strength of tip	Each use (after sterilization)
Interstitial equipment		
Source-holding needles	Straightness	Each use
	Patency	Each use
	Integrity	Each use
	Sharpness	Each use
	Bevel	Each use
	Diameter	Each use
	Length	Each use
	Connection (HDR)	Each use
	Collar (template)	Each use
	Funnel integrity (if attached)	Each use

Table 8.1. *(Continued)*

Article	Feature to be tested	Frequency
Interstitial equipment		
Catheter-inserting needles	Straightness	Each use
	Sharpness	Each use
	Bevel	Each use
	Diameter	Each use
Catheters	Integrity	Each use
	Diameter	Each use
	Length	Each use
Plastic buttons	Fit snugly yet slide	Each use
	Do not narrow catheter	Each use
Metal buttons	Slide onto catheters	Each use
Catheters with buttons attached	Buttons firmly attached	Just before use
Templates	Hole placement	Acceptance
	Hole angulation	Acceptance
	Needle guidance	Acceptance
	Needle fixation	After each use
	Obturator fit	Each use
	Obturator screw function	Each use
Ultrasound templates	Rotational alignment	Semiannually
	Scaling	Semiannually
Surface applicators		
Eye plaques		
Carrier	Holes cut cleanly	Just before loading
	Sources seat properly	During loading
Gold backing	Clean and smooth	Before construction
	Holes clear	After loading
Skin applicators	Thickness	Acceptance
	Source position	Acceptance
	Source fixation	Each use

CHAPTER 9

QUALITY MANAGEMENT FOR DOSIMETRIC TREATMENT PLANNING

'Treatment planning' in brachytherapy, as in external beam radiotherapy, means different things to different persons. Often the term substitutes as the equivalent of dose calculation. That forms a small, but important, part of treatment planning. Planning a brachytherapy procedure requires many decisions, and all too often a physician or physicist approaches planning with a fixed idea, bypassing many of the decisions and closing options that might better treat the patient.

The treatment planning process follows several phases. The timing of the progression between these phases and the actual insertion procedure varies with the particular procedure. A common time course follows:

(1) *Preprocedure imaging to determine a preliminary target*. While these images would be best made with the patient in the exact position as for the brachytherapy procedure, this often proves infeasible, and these images serve as interim guides to the target volume. In the absence of images, information from physical examination provides information on the size and extent of the target volume.

(2) *Preprocedure treatment planning, dose calculations and optimization*. Based on the images obtained above, this step entails deciding on the approach to the target, including:

- type of application—interstitial, intracavitary or surface;
- type of appliance—based on the type of applicator and anatomy of the patient. An example of this decision includes selecting between using needles to carry sources for an interstitial implant or replacing the needles with catheters;
- selection of accessories—for example, templates for interstitial implants or retractors for intracavitary insertions;
- dose rate—high, middle or low dose rate, or pulsed dose rate;
- dose uniformity criteria—should the dose cover the target uniformly or would a gradient through the volume conform better to the biology of

the target? If the dose should be uniform, what form of optimization should be used to achieve that uniformity?

- applicator and source placement—based on the above, determine the location of the appliance and the source.

Calculations based on the approach generated via the above decisions determine the preliminary, and tentative, treatment plan.

(3) *Procedure.* In this step, the physician, with the assistance of the medical physicist, executes the planned treatment as best as possible.

(4) *Postprocedure imaging.* This imaging serves to localize the sources and imagable anatomy for input into the dose calculation system.

(5) *Postprocedure dose calculation.* The dose calculation in step (2) reflected the ideal application. The calculations in this step incorporate the perturbations in the plan encountered during execution. The management of the procedure (for example, for HDR brachytherapy cases, or LDR cases where the source order to the manufacturer comes after the calculation step) may allow some flexibility in adjusting for deviations from the plan through reoptimization of the source distribution. After this recalculation of the dose distribution, the treatment duration (treatment time) calculation for LDR applications follows.

Not all procedures follow this programme, and many procedures either skip steps or invert some of the order. Not uncommonly, both localization and dose calculation occur twice, once in planning and once following insertion of the appliance.

Dosimetry for brachytherapy treatment planning consists of two distinct phases: localization and dose calculation. Both processes require acceptance testing and periodic re-evaluation. An important part of the acceptance testing is to familiarize the operator with the operation and peculiarities of the system. One important tool in this process is entry of deliberate errors. The operator should know how errors in each input affect the output, and how robust the algorithms are to erroneous peripheral parameters. This practice also allows the operator to learn methods of recovery, particularly after landing in unintended parts of the programme, and how such jumps might influence the resulting plan.

9.1. LOCALIZATION

Quality considerations for imaging and localization as applied to brachytherapy extend beyond the confines of the discussion below. For a more complete discussion on imaging in brachytherapy, see Paliwal *et al* (1997). Quality assurance procedures for diagnostic imaging systems can be found in NCRP (1988) and will not be repeated here.

The first requirement is that the imaging system produces high quality images. Contrary to the oft-expressed opinion that radiotherapy imaging requires less from a system than diagnostic imaging, brachytherapy localization demands the following abilities:

- *To resolve small objects.* Many of the sources used in brachytherapy present image cross sections of less than a millimetre. While resolving a single object of that size on a radiograph in a homogeneous background poses little problem, in a patient with anatomy evident, finding the objects becomes challenging. Other modalities, such as CT, face increased difficulties.
- *To present precise geometric information.* Not only must the system image small objects, but the position of the objects in space must be known to a high degree of accuracy. Small errors in the locations of sources or targets produce large deviations in the dose distribution. The errors in positions can come from image distortion or simply poor visibility.
- *To locate low contrast boundaries.* Not only must radiotherapy images locate a tumour, as must diagnostic images, but those for treatment planning require solid information as to the exact boundary of the tumour (something usually unnecessary in diagnosis). These boundaries often show little subject contrast between the healthy and target tissue.

Thus, imaging in radiotherapy requires top quality equipment in optimal condition.

The preprocedure imaging must be able to differentiate between the target and normal tissue structures. As such, this imaging usually uses CT, MRI or US, since radiographs seldom contain sufficient soft-tissue contrast. Increasingly, computerized dose calculation systems utilize such axial image sets with brachytherapy planning. Unfortunately, CT in the pelvis, a common site for much of brachytherapy, differentiates poorly between many of the structures of interest, particularly between the uterus and tumours in that organ, and even the boundaries of the prostate. MRI, on the other hand, provides good differentiation between tissues and excellent resolution, but proves considerably more expensive, often requires a long wait between the appointment and the scan, and limits patient positioning due to the narrow diameter of the scanning tunnel. The first two problems are becoming less serious with the increase in the numbers of units in the field, and open-field units (which, currently, are still a rarity) will overcome the last. Ultrasound for treatment planning became more common with the popularity of US guided prostate plans, but remains underutilized for gynaecological applications. Very few patients with cancer of the uterine cervix receive any preprocedural imaging as part of their treatment planning. Recognizing that, because of limited access to imaging equipment or control over that equipment during the preprocedure imaging phase, these images often represent poorly the target and patient as encountered at the time of the procedure. The recommendations below for sectional localization help assure conformance between these images and the patient at the procedure—the quality control.

9.1.1. Radiographic localization

9.1.1.1. Acceptance testing (quality assurance). Most brachytherapy procedures use radiographic localization at some point, possibly as the primary data source for dosimetry or sometimes just for verification. Either way, the images must provide accurate, unambiguous geometric information. Acceptance testing for the localization system, other than the routine re-evaluations typical for diagnostic radiography equipment, includes geometric tests akin to those for linear accelerators. The tests should be repeated approximately semiannually, and include evaluations of:

Coincidence of the cross hairs with the designated axis. X-ray beams from diagnostic radiographic units have no particular ray more central than any other. The collimator may impose some symmetry to the beam, but that symmetry never plays a role in localization. However, the beams must have an indicated ray, usually marked with a radiopaque cross hair, to relate the beam's geometry to that of other beams. The particular unit may have some preferred ray, for example, that relates to the centre of a fixed film holder or, for a simulator, the isocentre. Regardless of the equipment design, the cross hair forms a beam axis, and this axis must coincide with the intended and supposed axis. If the light-localization field is used for setting the beam, the projected shadow of the cross hair must coincide with the radiographic image to within 1 mm. If the image of the cross hair is expected to intersect a different beam's axis, this must be tested and happen to within 1 mm.

Angular accuracy. Each of the methods for imaging the brachytherapy application, except general triangulation with markers on a frame, relies on knowledge of the x-ray beam orientation with a high accuracy (Niroomand-Rad *et al* 1987, Niroomand-Rad and Thomadsen 1990). All of the reconstructed methods in commercial treatment planning systems assume normal incidence of the beam axis to the film. In addition to producing grid cutoff, an angle between the film and the incident beam causes the magnification to change across the film.

Evaluating the images for accuracy (providing quality assurance) can simply use some blocks of Styrofoam™ with lead shot imbedded at known locations. A simple square pattern 10 cm on a side in two planes 7 to 10 cm apart (for example, on opposite sides of the Styrofoam block) provides a convenient input pattern. To test the accuracy of the reconstruction procedure, regardless of the type of localization (orthogonal pair, stereo shift etc), image this phantom and enter the images into the planning computer. It may be necessary to pretend the points represent sources or anatomical points in the patient. If the program handles the two types of input differently, enter the points under both inputs. The output of all modern planning systems lists the coordinates of the sources or anatomical points. Compare the coordinates with the known positions of the lead shot. The computerized coordinates probably contain a constant shift, but that should be in a known direction and by a predictable amount. The reconstruction should duplicate the coordinates of the points to within 2 mm.

9.1.1.2. Checks per case (quality control). The information necessary to interpret the localization films depends on the reconstruction method. While each method applies triangulation to determine the coordinates of objects in the patient, they use different fixed parameters to deduce these coordinates.

- *Orthogonal pair.* This method uses two radiographs with their axes oriented at right angles to each other. The lines connecting the focal spot and image of an object for the two radiographs intersect at the object. By knowing the coordinates of the focal spots and of the images in the object, the coordinate of the object can be inferred. The coordinates of the focal spot and the film usually come with the assumption that each focal spot lies on a coordinate axis (indicated by the cross hair), and from knowing two of the following: the focus–film distance (FFD), the focus–axis intersection distance (FAD), the magnification and the intersection (M). While many units limit those parameters that vary (such as C-arm units, which fix the FFD, or simulators, which may fix the FAD) very few systems fix the entire geometry (for example, the FAD may vary with a C-arm and the FFD with a simulator). As a check on the consistency of the measured values, two fiducial markers in a plate below the x-ray collimators project onto the film with a magnification

$$M_f = \frac{s_f}{s_c} \tag{9.1}$$

where

$M_f =$ the magnification of the markers in the film plane,
$s_f =$ the separation of the marker images on the film, and
$s_c =$ the physical separation of the markers at the bottom of the collimators.

The magnification from equation (9.1) should agree with that calculated by

$$M_c = \frac{\text{focus–film distance}}{\text{focus–fiducial distance}} \tag{9.2}$$

to within 2%. After verifying that the fiducial marks produce the expected results during their acceptance testing, in the absence of changes in the fiducial markers, larger differences between M_f and M_c indicate a mistaken FFD. Simulators often contain fiducial markers for determining field sizes for external-beam treatment, and such prove excellent for this test. (Some simulators permit variations in the FAD, in which case this test indicates only that the ratio between the FAD and the FFD differs from the expected without specification of which setting produced the change.)

Neither M_f nor M_c corresponds to the magnification used in the calculations. That quantity, M, relates the magnification of an object located at the intersection of the axes. Knowing the FFD and FAD determines M as

$$M = \text{FFD}/\text{FAD}. \tag{9.3}$$

If either of these distances are uncertain, the magnification can be determined using a magnification ring. The ring, with a known diameter (usually about 5 cm), simplifies determining magnification since the largest dimension of its image in the film plane corresponds to a projection of its diameter, as

$$M = \frac{\text{largest dimension of the ring}}{\text{actual diameter of the ring}}. \tag{9.4}$$

However, this procedure only gives the magnification at the distance of the ring. Since the magnification required in the reconstruction of object coordinates corresponds to that at the intersection of the beam axes, the ring position for one of the beams (for example, a lateral) falls in the central ray of that for the complementary beam (in this example, the anterioposterior). Radiopaque cross hairs usually indicate the central ray and help guide the placement of the magnification ring. Even if the distances are known, a magnification ring provides a redundant check on the data entered into the reconstruction algorithm.

The angle between the beams forms another important variable. Verification of the angle can use a fiducial protractor as described below under 'variable angle'.

- *Stereo shift.* The stereo-shift method uses two vantage points separated by a translational shift, to provide the parallax for establishing the coordinates of points in the patient. Fitzgerald and Mauderli (1975) and Sharma *et al* (1982) discuss the uncertainties for this procedure and requirements to provide accuracies necessary for clinical practice. The accuracy of reconstruction based on a stereo shift increases with the divergence of the two x-ray beams and their separation. Thus, the desirable conditions find long focus–film distances (assuming that the distance from the patient to the film remains invariant) and large shifts. The uncertainty in the z coordinate in figure 9.1 decreases also with increasing FFD. Fitzgerald and Mauderli recommend FFD $\geq$ 1 m and $s \geq \text{FFD}/2$.

  A jig, as shown in figure 9.1, assists in setting the prescribed shift between the two radiographic exposures. The known separation of the markers in the jig falling at a known distance above the film allows confirmation of the FFD as:

$$\text{FFD} = \text{OFD}\frac{i}{i - r} \tag{9.5}$$

  where

  OFD = the jig to film distance
  $i =$ the separation between the images of the markers on the film, and
  $r =$ the actual separation between the markers on the jig.

- *Variable angle.* The reconstruction methods included in this category form a hybrid between the orthogonal pair and the stereo shift. Figure 9.2 illustrates

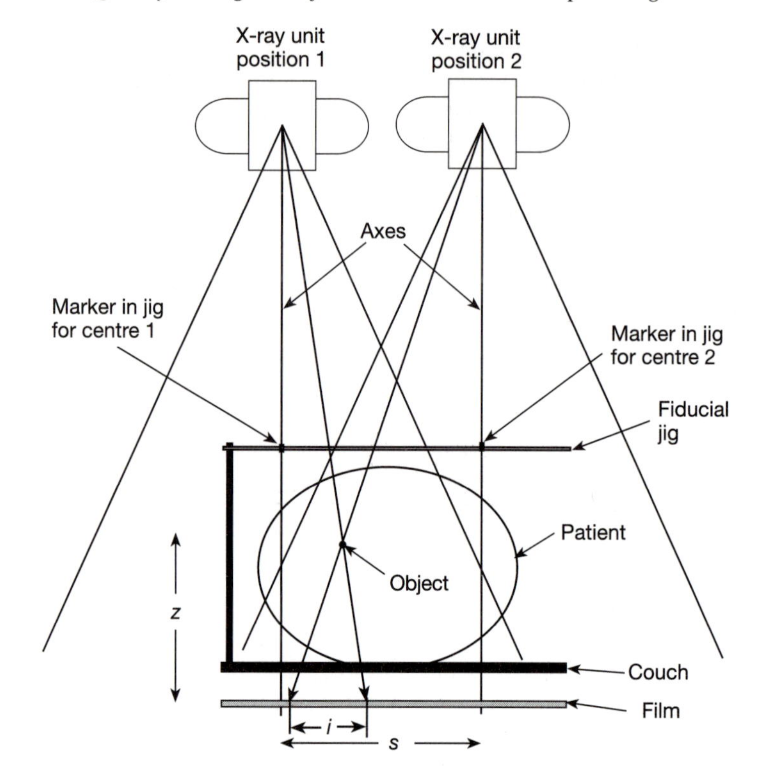

Figure 9.1. *A schematic diagram showing the use of a fiducial jig to assist in setting the x-ray units for a set of stereo-shift images.*

the basics of the geometry. Because this technique almost always finds application on a simulator, the unit most likely contains fiducials in a plate at the bottom of the collimators with which to check the value of the FFD. The angle between the beams remains the major variable, usually undocumented in most systems, although with the very rapid proliferation of record and verify systems, this, along with the other reconstruction data, is captured at an increasing number of facilities. One way to record the angle of each beam on the radiograph uses a fiducial protractor as shown in figure 9.3. The protractor carries lead indicators of the beam entry angle and slides into brackets attached to rails on the couch. This configuration allows alignment of the protractor origin with the isocentre using the side lasers (necessary since the table height varies with the thickness of the patient) and movement of the protractor to a location on the radiograph not obscuring any important structures. By having the protractor continue under the patient forming a circle (impossible on some units), a deviation from a reading of the identical angle on the beam exit indicates an error in the alignment of the protractor

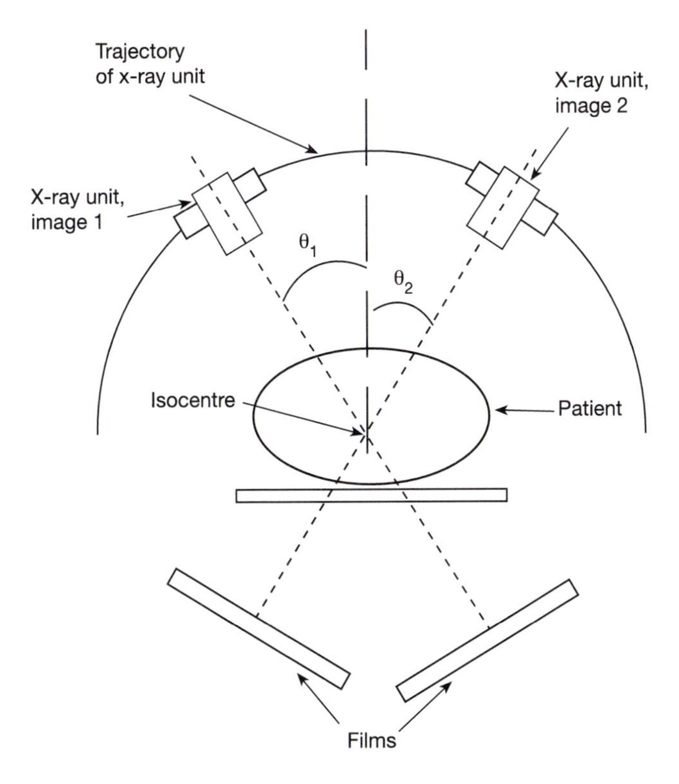

Figure 9.2. *The variable-angle technique for localization.*

with the beam isocentre, and thereby possibly an error in the alignment of the patient compared with that intended.

- *General triangulation.* This method of reconstruction assumes no *a priori* relationship between the two vantage points (focal spots). Instead, a frame around the patient contains fiducials on all four sides (see figure 9.4). From the images of the fiducials in the frame the computer program backprojects the positions of the focal spots relative to the centre of the frame, which acts as the origin for the reconstruction. The advantage of this method is that the positioning of the x-ray sources becomes arbitrary, requiring neither care nor quality control. After acceptance testing of the frame and its fiducials, the method calls for little quality control altogether. As a quality assurance procedure, an additional marker can be inserted near a corner of the frame in a place unlikely for the patient to occupy. During digitization, comparison of the reconstructed coordinates of the marker with those measured during acceptance testing gives a measure of the uncertainty for the procedure, with large differences indicating potential problems or errors in the localization procedures.

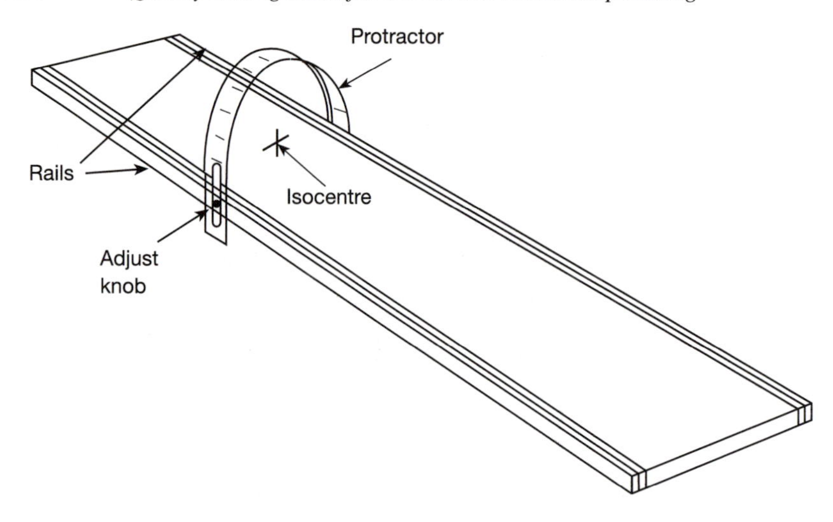

Figure 9.3. *A protractor containing radiopaque markers to document the angles used for reconstruction using the variable-angle technique.*

9.1.2. Sectional localization

While sectional imaging provides the most reliable information on patient anatomy for preprocedural planning, for postprocedural imaging, these modalities pose serious problems.

Most sectional imaging modalities provide excellent spatial resolution in the planes originally imaged, usually transverse, or axial, planes. Ultrasound resolution depends on the distance away from the focal distance of the transducer, with the resolution degrading either closer to the probe or further away. Yet for most sectional imaging, reconstruction in the third dimension, the one that separates the image planes, provides greatly inferior resolution. The resolution in this direction remains limited by the separation of the original image planes. Thus, for example, identifying the tip of an implant needle on CT images with 0.5 cm separation gives an uncertainty in the axial position of that tip of ± 0.25 cm. This uncertainty in the tip position translates directly into corresponding uncertainties in the positions of sources in the needle. The same uncertainties hold for identifying seed-type sources, but notice that if the axes of two seeds fall parallel to the image planes, and each seed shows only on adjacent planes, the seeds could fall almost 1 cm apart or almost touching. Were these seeds near the urethra, the two possible positions produce markedly different expected outcomes. Improvement in the resolution in the axial direction comes only with decreased plane separation.

MRI gives the possibility of making the images in planes other than transverse. When combined with the axial study, this may increase the resolution between the axial planes, but with the same limitations due to spacing of these

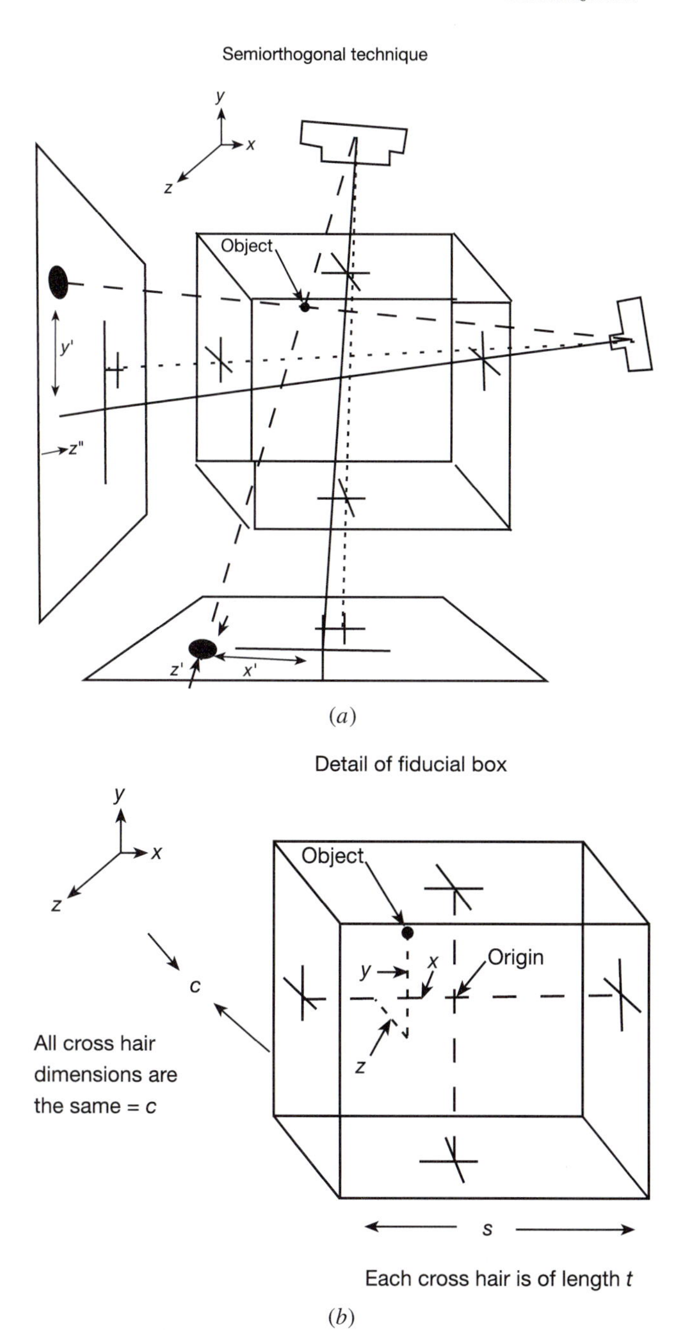

Figure 9.4. *A diagram showing the concept of general triangulation for localization.*

additional planes. Unfortunately, few computerized planning systems combine images taken in different orientations.

Appliance-produced artifacts become a serious problem in most settings. While CT- and MRI-compatible gynaecological applicators have become available (for a considerable price), the usual plastic needles compatible with these modalities occasionally break in a patient or prove too flaccid to maintain the correct geometry. Recently titanium and tungsten needles have become available that cause no artifact on the images. With US, any needle produces artifacts that confuse the image and can hide distal needles.

Once overcoming all of the above problems, to use the sectional images for dose calculations requires some further quality control procedures, probably either at quarterly or annual intervals:

(1) *Evaluation of image distortion.* A distortion of the images causes no harm if the display screen simply distorts all parts of everything shown. (This is *not* true of the printouts, discussed elsewhere.) The 'everything' includes the brachytherapy appliance and dose distributions. Problems arise if the image data suffer different distortion from *any* other data displayed. This situation can result in inappropriate positioning of sources or treatment times. In general, the most common image distortions take the form of:

- *Scale distortion*, where the image display fails to appear with the same magnification as the rest of the display. This often occurs due to misread file headers, and probably affects the axial magnification as well as that in one or both of the planar coordinates.
- *Scale nonlinearity*, where the shape of the image display fails to conform to the object imaged. An example of this type of distortion is a 'pin cushion' effect, where the image of a Cartesian grid bulges in the middle. This type of distortion usually appears in the original image set as well as that transferred into the treatment planning computer.

Assessing image distortion entails imaging a phantom with small, relatively high contrast objects at known locations at various parts of several imaging planes. Transferring the images into the treatment-planning computer, the determined coordinates of the objects compared to the coordinates in the phantom establish the deviations. In general, deviations greater than 2 mm give cause for concern regarding the accuracy of resulting dose distributions.

(2) *Coordination of coordinate systems.* The image coordinate system may not correspond with that of the appliance initially, particularly if the latter originated from a different image set (such as radiographs). Ensuring that the appliance, and thereby the sources end up in the correct location in the patient requires coordination of all coordinate systems. Unfortunately, this process usually happens individually for each patient. The coordination requires an identifiable object in each data set if both sets follow the same orientation or at least three objects if the relationship between the orientations remains unknown.

The problem may present itself in reverse, developing a treatment plan based on an image data set, and assuring the insertion of the applicators matches that plan. In this case, the image set needs unambiguous information to guide the applicator insertion. Consider the example of a temporary, template-based implant as a boost for treatment of a prostate following external-beam therapy. (Chapter 3 considers this example in detail.) A marker corresponding to a tattoo on the perineum along with the anus give reference points for establishing the insertion points and rotational orientation for the template. A small, contrast-filled bulb in the Foley catheter provides a target for a guide needle visible under fluoroscopy at the time of the implant, to set the angular direction of the needles.

9.1.3. Ultrasound localization

Ultrasound images pose a different, and additional, problem when used with treatment planning. Unlike CT or MRI, most US scans have no particular geometric orientation. Rather, the operator obtains them freehand. To make use of US images requires making a frame that constrains the probe to image in a single plane at a time, and translate only perpendicularly to the imaging planes. (Ironically, very old and obsolete B-mode units performed this exact function. Not all progress moves forward.) The exception to this situation is transrectal US for prostate brachytherapy (or transvaginal US for gynaecological implants). Transrectal US presents the problem of defining the orientation of the image planes, as discussed in the section on templates in chapter 8. While the US/template system *assumes* a particular geometry, this needs testing and verification, as in chapter 8. As mentioned above, an additional problem encountered using US for dosimetry is the artifacts in the images induced by needles or sources. At the time of writing, no system can filter these artifacts. The range of image information also limits some of the modality's usefulness in planning. As an example, the short range of the transrectal ultrasound image provides no information on the position of the pelvic arch and its possible interference with the needle trajectories.

Even the sectional images of transrectal US present problems for reconstruction. Each source or needle implanted and in the image slice produces numerous artifacts, and attenuates the signal masking more distal features. While useful for planning imaging, as a reconstruction modality following implantation, ultrasound remains unworkable.

Other than the normal routine quality management for ultrasound units, and those tests discussed in chapter 8, the only additional check evaluates the relative positions of the axial and sagittal planes on units that allow automatic switching between. Some units rotate around a single axis, while others use two transducers, and thus have a shift between the locations of the cuts. Regardless of the nature of the unit, the relationship between the planes must be known.

Currently, several companies are working towards three-dimensional US imagers. With the advent of such units, the question of quality management for brachytherapy treatment planning will require revisitation.

9.1.4. *Source simulation*

The localization procedures that do not image the sources proper simulate the sources by non-radioactive markers[1]. The markers only perform a useful function if they indicate the same locations as the sources will occupy later. To satisfy this criterion usually requires two properties of the markers.

9.1.4.1. Correct spacing of the markers in a train. The markers need not physically occupy the same locations as the sources, as long as the markers unambiguously indicate the source positions. For example, HDR source dwell positions fall at 2.5 mm intervals along a catheter. The markers usually indicate only every fourth dwell position with indicators at 1 cm intervals. By entering the centimetre markings, the computer calculates the intervening source positions by interpolation between the markers. In most cases, this procedure proves adequate. However, large curvatures can produce marked differences between the straight-line interpolations and the curved path of the source. Quarterly, the separation of the indicators in the marker trains should be checked for positional accuracy, and the train replaced if deviations from the specified separation exceed 1 mm. Markers in nylon ribbons require more frequent testing, possibly before each use, because any fluid in the ribbon reduces the friction holding the indicators in place. The indicators in such markers often slide out of place by several millimetres over a year's period, even without fluid entering the ribbon. Indicators in the form of metal beads crimped onto a cable tend to slide less often than indicators in nylon ribbon, but such movement still happens. In addition, the cables tend to kink, often due to rushed insertion into the appliance, which causes a shortening of the interval at the kink. Markers for LDR caesium sources must duplicate the geometry of the source. An exception to this rule occurs with sources where the active lengths fall asymmetrically in the physical jacket (usually due to the extra space taken by the eyelet). In these cases, the treatment-planning computer may automatically (with no choice) centre the active length in the entered physical length. To end with the source in the correct location, the marker needs to provide a false indication of the ends of the source symmetrically positioned around where the active material will truly reside. An example of such a marker would be a plastic rod of the source's true physical length, with small lead indicators embedded at the tip on one end, but at the fictional, symmetrically positioned location on the other (see figure 9.5). Using such an indicator keeps the active lengths in the correct positions with several sources in a row—provided the sources are loaded with the right orientations.

[1] In this text, the term 'marker' applies to objects that simulate the sources for the purpose of treatment-planning imaging, with an 'indicator' being that part of a marker actually simulating a source location, while the term 'dummy' refers to nonradioactive objects physically replacing sources for the purpose of performing some nonimaging tests. For remote afterloaders, cables that travel through the transfer tubes and appliances preceding the source cable go by the term 'check cable'.

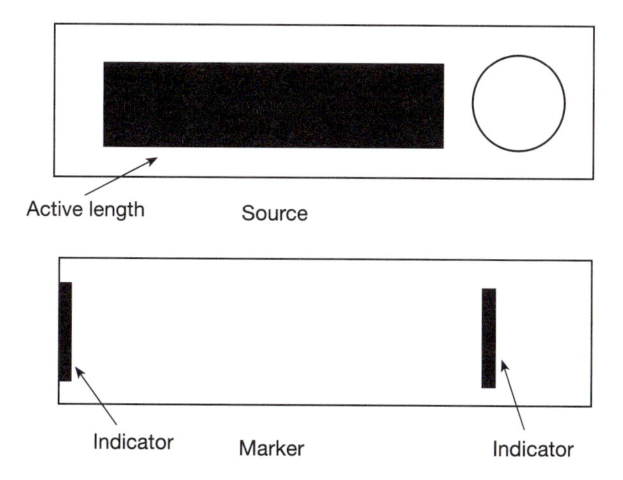

Figure 9.5. *A marker for an asymmetric tube-type source to produce the correct positioning of sources in a train.*

9.1.4.2. Correct positioning along the appliance tracks. Placement of manually loaded sources usually inserts the source or source train to the end of the treatment appliance. For such cases the markers must also reside at the end of the track. The important aspect of the marker to mimic the sources lies in the length of the leader before the indicator to make the centre of the indicator coincide with the source. Figure 9.6 illustrates this. For ^{192}Ir sources in a nylon ribbon treated as point sources in a treatment planning system, the indicators need not have the same length as the sources, but the combination of the leader and half the source length must equal the marker's leader length and half the length of an indicator. Preparation of the source trains for insertion includes checking that the length of the leader corresponds to that consistent with the desired position for the sources.

Several remote afterloaders also follow the custom of inserting the source train to the end of the applicator. However, some afterloaders, and some manually afterloading appliances, such as Fletcher–Suit ovoids, position the sources relative to some part of the system on the open end of the source track. The Nucletron microSelectron HDR remote afterloader sends its single ^{192}Ir source along a treatment catheter to a distance specified with respect to a sensor in the head of the unit. In the Fletcher–Suit ovoid, the source carrier snaps a pin into a hole in the ovoid handle, fixing the position of the bucket that holds the source. For any system operating in such a fixed-length mode, the markers must have some method of positioning the indicators to mimic the source. In the high dose rate unit, a collar on the source train abuts the transfer tube orifice, suspending the markers in a standard position. Part of the daily tests for the unit entails assuring conformance between the markers and the source dwell positions. Any depart-

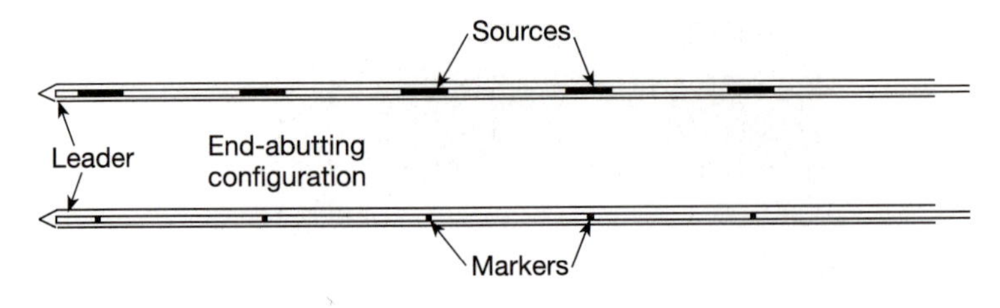

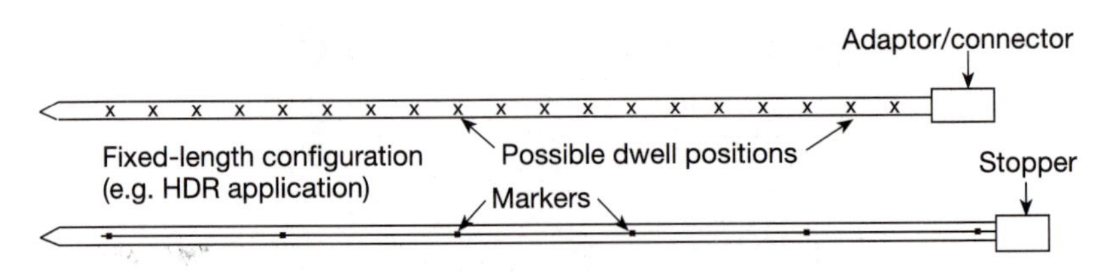

Figure 9.6. *Positioning x-ray markers properly in a needle. The upper figure shows a manually loaded situation where the source train abuts the end of the needle. The leader of the markers and half the length of a marker seed must equal that for the sources. The lower figure shows a fixed length application. In this case, the marker distance from the end must equal the value for the source. Whether the sources go to the end of the needle or to a fixed position, the markers must duplicate the source positions. Using the wrong marker train for the application (for example, an end-abutting train for a fixed position source placement) can result in serious errors in source position reconstruction.*

ment that performs treatments with both end-residing and fixed distance systems must keep the markers for the systems clearly indicated and separate to avoid misuse. Seldom do the markers for one system follow a geometry appropriate for the other.

Williamson (1991) and Ezzell (1999) note differences between the indicators and source positions when the markers function differently from the source. Williamson observed that cable-connected LDR sources for a remote afterloader occupy vaginal ovoids at an angle (see figure 9.7). Using markers in the customary buckets, or using no markers and assuming that the source falls centred in the ovoid cavity lead to erroneous dose distributions. Ezzell observed that the indicators on the marker cable in a vaginal ring applicator, because the cable supports them on both sides, lie at different positions than the actual HDR source, which only connects to the drive cable on one side (see figure 9.8). These examples illustrate the importance of simulating the source as closely as possible with the localization markers.

Figure 9.7. *An ovoid containing a source from an LDR remote afterloader. Notice that the cable connection prevents the source from assuming the normally expected position.*

9.2. DOSE CALCULATION

The calculation of dose proceeds through several steps, each with its own considerations with respect to quality. Failure in any of the steps results in erroneous doses from the planning computer, and potentially dangerous errors in the patient treatments. The discussion below centres on typical treatment planning computer configurations common at the time of writing. The systems are changing very rapidly, however. Yet, the considerations probably will serve as guidelines even for future systems, although the details may have to be adapted. For further guidance on quality management of dose-calculation computers, Curran and Starkschall (1991) present a good discussion for a comprehensive program. The report of Task Groups 40 and 53 of the American Association of Physicists in Medicine (AAPM TG40 1994, AAPM TG53 1998) also presents a recommended program for dose calculation quality management.

Obviously, dose calculations performed prior to the procedure[1] use idealized source positions that contain no uncertainty. In those cases, the quality considerations in the following discussion begin with the section on 'Dose calcu-

[1] The vernacular often refers to dosimetry performed before the procedure as a 'preplan'. The term 'preplan' by construction is redundant (and not in the constructive sense as discussed in chapter 1) since any 'plan' occurs before the procedure. The term for dose distributions after the procedure, 'postplans', is illogical, since one does not plan for something after it occurs, except possibly politicians.

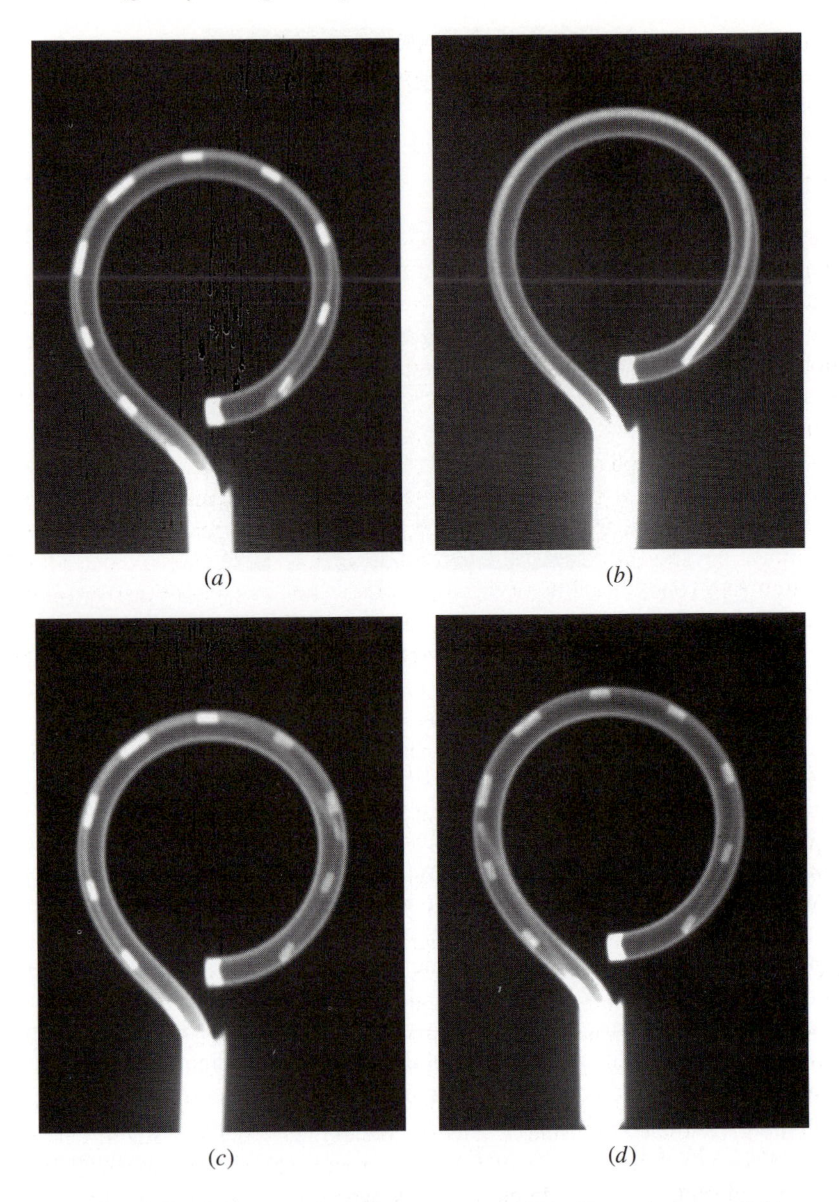

Figure 9.8. *Illustration of problems using a cable with markers in a tight bend. (a) The marker with indicators at every fourth dwell position based on the distance along the marker. (b) The source at the first dwell position. (c) Superimposed images of the markers and the source at the ninth dwell position. Some misregistration can be seen. (d) Source at the 29th dwell position. Significant differences exist between the marker (second from the stem clockwise) and the source position. (Figure courtesy of G Ezzell.)*

lation' (section 9.2.3). A major consideration for prevention of error in delivering the planned doses becomes placing the sources in the planned locations, a topic discussed in several previous chapters.

9.2.1. Digitization

Regardless of the nature of the images used for localization, the information from the images enters the computer through a digitization procedure. The most automated, although not necessarily the most error free method reads the images directly into the treatment planning computer, usually by network transfer of sectional images directly from the imaging computer, with software identification of sources or dummy markers. Often, the planning system also contours anatomic structures of interest with minor input from the operator. Currently, such systems find the greatest application in evaluation of the dose from permanent implants for prostate cancer, imaged with CT approximately one month after the procedure. Digitization by this method contains two major uncertainties: the inherent positional uncertainty discussed above in the section on 'Sectional localization' (section 9.1.2) and the ability of the algorithm to 'find' sources or markers in the image and correctly assign the position of the centre of the source. The latter operation can be influenced by the image content near the source and by artifacts in the image, possibly caused by other sources or the source itself. A step removed from this automated approach finds the operator finding the sources in the image and digitizing them on the screen. Reference to radiographs taken just before or after the sectional images sometimes assists in selecting the most appropriate slice to assign a source that appears on more than one image, or identify images that may represent two sources in close proximity. The manual entry adds the uncertainty of positioning the pointer on the computer screen, but that uncertainty seldom comes close in magnitude to others.

Digitizing information from radiographs (for example, from an orthogonal pair) currently can use any of many technologies: backlit digitizing tablets using the hardcopy films, computer-screen cursor using scanned copies of the films, or computer-screen cursor using captured fluoroscopy images. All these technologies provide digitization uncertainties an order of magnitude better than the operator's ability to point the cursor. Still, the systems require periodic testing to assure their calibration. The calibration check can use a radiograph of the Styrofoam phantom such as described above in section 9.1.1.1, except with a larger square formed by the lead shot. A square 25 cm on a side works well, with one marker in the centre. Laying the phantom with the shot directly in a piece of film and exposing it appropriately provides an image without magnification. Digitizing this image tests the accuracy of the digitization system over a large area. The digitization should reproduce the known coordinate of each of the markers with respect to the centre marker to within about 0.5 mm, that is, the limit of the human aspect of the input. This test probably should be performed daily (as recommended by TG 40), not because of a large likelihood of erroneous digitization (modern digitization seldom fails except as outright dysfunction), but because periodic testing probably will miss detecting changes before they affect patient dosimetry, and such changes carry extremely serious consequences.

9.2.2. *Geometric reconstruction*

Imaging and reconstruction of the Styrofoam phantom such as described in section 9.1.1.1 should give the relative coordinates of the markers to within 1 mm. (The positions of the indicators should be confirmed after construction of the phantom without assuming that the indicators ended at the desired locations, for errors of the order of 1 mm in indicator placement occur easily with Styrofoam.) This test probably need be performed only during acceptance testing of the system. An exception to the limit on necessary repetition of this test happens following entry into a file in the computer that contains any of the parameters used for the reconstruction. During work on the file, data for the reconstruction could accidentally be changed. Repeating the test verifies the correct operation of the reconstruction.

9.2.3. *Dose calculation*

Most obviously, errors in the dose calculational algorithm or any of its factors produce consequential errors in the calculated dose distribution. While one would expect that the algorithm and its factors would be easy to check, unfortunately that turns out not to be the case. Confusion occurs mostly for two reasons:

(1) Many computer systems take the input parameters from the operator and calculate a 'lookup' table. Unfortunately, the various commercial units that create these tables include different combinations of the variables. The manuals frequently become vague concerning the exact combinations. Sometimes the computer deletes the input data after constructing the table.
(2) Some computers use algorithms written long ago before the common inclusion of some factors. As a result, the 'new' factors had to be rolled into the old. This problem became extremely prevalent with the advent of the formalism of Task Group 43 of the American Association of Physicists in Medicine (AAPM TG43 1995).

Instead of using the standard equations for dose calculations, some manufacturers fit the dose calculated by the equations to other functions, such as a polynomial. Such a switch in the algorithm obviates inspection to detect errors in the function.

The change to TG43 also brought a new wave of errors in factors used, particularly the misconception of equivalence between the radial dose function in the TG43 formalism and the tissue-effect coefficient (of which the Meisberger polynomial is an example) in the old. Based on equation (9.6) for the old formalism and equation (9.7) for TG43, equation (9.8) gives the relation between these two quantities, and table 9.1 the value for the conversion for some relevant sources.

Table 9.1. *Conversion factors to give $T_{med}(r)$ from $g(r)$ for some sources.*

Isotope	f_{med}	$\frac{S}{A}\frac{\Lambda}{\Gamma f_{med}}$
^{125}I model 6711	0.886 cGy R^{-1}	0.870
^{125}I model 6702	0.885 cGy R^{-1}	0.920
^{103}Pd	0.891 cGy R^{-1}	0.725

Figure 9.9 illustrates the geometric parameters for the following equations. For the older, conventional formalism, the dose rate follows

$$\dot{D}_{in\,medium}(r) = A\Gamma f_{med} T_{med}(r) G(r,\theta)\phi'(r,\theta) \tag{9.6}$$

where

r = the distance from the centre of the source to the calculational point,

$\dot{D}(r)$ = the dose rate at r under the conditions in the subscript,

A = the activity of the source,

Γ = the exposure rate constant,

f_{med} = the exposure-to-dose conversion factor,

$G(r,\theta)$ = the effects of geometry,

$\phi'(r,\theta)$ = a function that accounts for anisotropy and

$T_{med}(r)$ = the tissue-effect factor, where

$T_{med}(r)$ = $\frac{\dot{D}_{in\,medium}(r)}{\dot{D}_{in\,air}(r)}$.

For pointlike sources (including linear sources for r greater than twice the active length), $G(r,\theta)$ includes the inverse square relationship, while for a linear source,

$$G(r,\theta) = \frac{\beta}{Lh}$$

where β is shown in figure 9.9, L = the source length and h = the perpendicular distance from the source axis.

The factor $\phi'(r,\theta)$ accounts for anisotropy due to the encapsulation, for example, accounted for using the Sievert integral. Since $\phi'(r,\theta)$ includes correction for the attenuation of the source encapsulation, factor Γ must also be corrected to the value appropriate for an unencapsulated source.

According to the TG43 formalism,

$$\dot{D}_{in\,medium}(r) = S_k\Lambda \frac{G(r,\theta)}{G(r=1,\theta=\pi/2)} g(r)\phi(r,\theta) \tag{9.7}$$

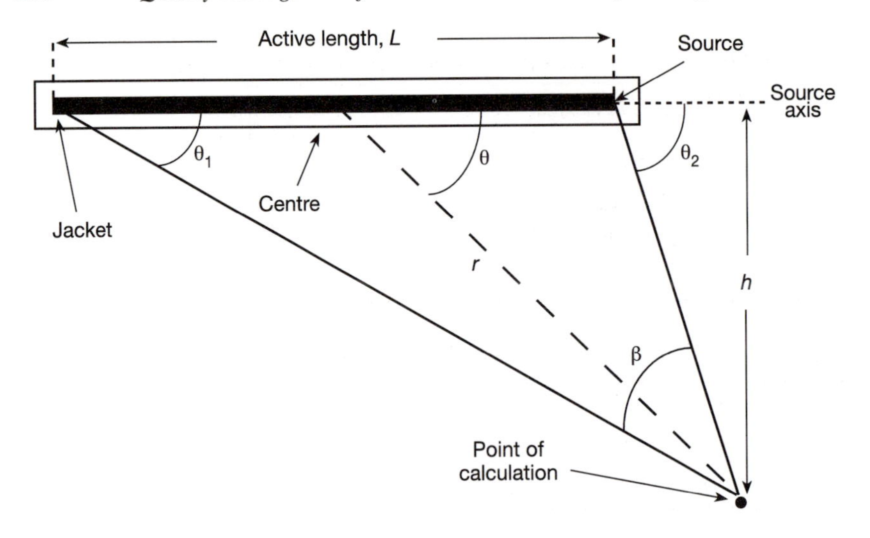

Figure 9.9. *The geometric parameters for dose calculations from a brachytherapy source.*

where

S_k = the source air kerma strength as assayed by the institution for each source,
Λ = the dose rate constant for the source type,
$G(r, \theta)$ = the geometry factor, excluding any effect from source encapsulation,
$\phi(r, \theta)$ = the anisotropy function and
$g(r)$ = the radial dose function, where

$$g(r) = \frac{\dot{D}_{in\,medium}(r)}{\dot{D}_{in\,medium}(r = 1\text{ cm})}$$

with the geometry function removed. The geometry function for a point source equals $1/r^2$, and for a line source β/Lh. The anisotropy function degenerates into the anisotropy constant for randomly oriented sources.

The radial dose function relates the dose at a distance r *on the perpendicular bisector* of the source to the dose at $r = 1$ cm, both in medium. In the old formalism, $T_{med}(r)$ related the dose at some distance r in the medium to the dose at the same distance in a small mass of medium just large enough to produce electronic equilibrium, surrounded by air (referred to as the dose 'in air', but not to air). In terms of $g(r)$, values for $T_{med}(r)$ become

$$T_{med}(r) = g(r)\frac{S}{A}\frac{\Lambda}{\Gamma f_{med}} \qquad \text{for a point source} \qquad (9.8a)$$

Table 9.2. *Values for the calculational parameters in the TG43 formalism for some selected sources.*

Source type	Λ (Gy h^{-1} U^{-1})	Anisotropy constant
^{125}I model 6711	0.0098	0.93
^{103}Pd model TheraSeed	0.0074[a]	0.90
^{192}Ir	0.0112	0.98

	Radial dose function coefficients					
Source type	a	b	c	d	e	f
^{125}I model 6711	1.014	1.227×10^{-1}	-1.730×10^{-1}	4.024×10^{-2}	-3.852×10^{-3}	1.343×10^{-4}
^{103}Pd model TheraSeed	1.629	-7.686×10^{-1}	1.543×10^{-1}	-2.272×10^{-2}	3.393×10^{-3}	-2.672×10^{-4}
^{192}Ir	0.9890	8.813×10^{-3}	3.518×10^{-3}	-1466×10^{-3}	9.244×10^{-5}	—

$g(r) = a + br + cr^2 + dr^3 + er^4 + fr^5$.

[a] These sources are just being calibrated by NIST at the time of writing, and this value may change as a result. The user should check with the manufacturer for literature giving the most recent value.

and

$$T_{med}(r) = g(r)\frac{S}{A}\frac{\Lambda}{\Gamma f_{med}}\frac{L(h = 1\text{ cm})}{(\beta \text{ for } h = 1\text{ cm})} \qquad \text{for a line source.} \tag{9.8b}$$

Care must be taken that the units match.

Table 9.2 gives current values of the parameters for dose calculations based on TG43 formalism and source strengths traceable to the National Institute of Standards and Technology in the United States for some sources common at the time of writing. Many new sources are coming to the marketplace, and the values in the table should only be used to test the computer calculation algorithm, and check the values for the actual sources listed. These values change periodically as research provides better information or for sources with strengths traceable to other national standard laboratories.

Because of the many types of ^{137}Cs source in use currently, users should check published dose tables for their particular sources. The only sources available at the time of writing are the Amersham type J sources. Table 9.3 gives the dose rates for these sources as a function of position.

Anisotropy function tables from AAPM TG43 for some sources appear as an appendix to this chapter.

Table 9.4 presents dose rates at various values for radial distance and polar angle for several sources. The dose rates at these points can serve as a test for the calculation performed in a computer. Figure 9.10 depicts a three-dimensional array of sources and calculation points. While the previous test evaluated the

Table 9.3. *Dose rate per unit source strength (cGy $h^{-1}U^{-1}$) for Amersham J-type ^{137}Cs tube sources (1992 design: vendor specifications). (From Williamson 1998a. Reproduced by permission of Elsevier, Amsterdam.)*

Distance along (cm)	Distance away (cm)						
	0.0	0.25	0.5	0.75	1.0	1.5	2.0
0.0	—	8.084	3.087	1.612	0.979	0.463	0.266
0.5	—	6.557	2.484	1.350	0.854	0.426	0.252
1.0	—	1.589	1.141	0.805	0.585	0.341	0.218
1.5	0.539	0.522	0.500	0.434	0.363	0.250	0.176
2.0	0.271	0.264	0.266	0.252	0.229	0.180	0.138
2.5	0.165	0.162	0.163	0.161	0.153	0.130	0.107
3.0	0.112	0.110	0.110	0.110	0.107	0.0966	0.0834
3.5	0.0806	0.0798	0.0791	0.0798	0.0789	0.0735	0.0658
4.0	0.0608	0.0604	0.0595	0.0602	0.0600	0.0573	0.0527
5.0	0.0380	0.0379	0.0372	0.0374	0.0376	0.0370	0.0352
6.0	0.0259	0.0258	0.0254	0.0254	0.0255	0.0254	0.0247
7.0	0.0186	0.0186	0.0183	0.0182	0.0183	0.0183	0.0180

Distance along (cm)	Distance away (cm)						
	2.5	3.0	3.5	4.0	5.0	6.0	7.0
0.0	0.171	0.119	0.0872	0.0664	0.0419	0.0286	0.0206
0.5	0.165	0.116	0.0854	0.0654	0.0415	0.0284	0.0205
1.0	0.149	0.108	0.0807	0.0625	0.0403	0.0278	0.0201
1.5	0.128	0.0958	0.0737	0.0581	0.0383	0.0268	0.0196
2.0	0.106	0.0829	0.0657	0.0530	0.0359	0.0256	0.0189
2.5	0.0868	0.0705	0.0576	0.0475	0.0332	0.0241	0.0181
3.0	0.0707	0.0594	0.0499	0.0421	0.0304	0.0225	0.0171
3.5	0.0577	0.0499	0.0430	0.0370	0.0276	0.0209	0.0161
4.0	0.0474	0.0420	0.0370	0.0325	0.0249	0.0193	0.0151
5.0	0.0329	0.0302	0.0275	0.0249	0.0201	0.0162	0.0130
6.0	0.0236	0.0222	0.0207	0.0192	0.0161	0.0134	0.0111
7.0	0.0175	0.0167	0.0159	0.0149	0.0130	0.0111	0.00945

To obtain cGy (mg Ra eq)$^{-1}$ h^{-1} multiply by 7.227.

algorithm and the values for the factors in the equation, this problem tests for the ability of the dose computation system to manage sources in space. Differences in dose rate of 2% may be found based on the actual algorithm used in the calculation. Greater differences warrant further investigations. Whether the discrepancy remains constant or varies with distance and angle provides clues as to which functions to examine. Williamson (1991) points out that when comparing

Table 9.4. *Dose rates per unit source strength (cGy h^{-1} U^{-1}) and doses (cGy U^{-1}) for a 5 d application per unit source strength for selected points and sources.*

Source type	$r = 3$ cm $\theta = 90°$	$r = 3$ cm $\theta = 45°$	$r = 2$ cm $\theta = 45°$	$r = 2$ cm $\theta = 30°$	$r = 5$ cm $\theta = 30°$	Dose at $r = 3$ cm, $\theta = 90°$ for 5 d
^{125}I model 6711	0.0688	0.0579	0.189	0.166	0.0107	8.022
^{103}Pd model TheraSeed	0.0255	0.0185	0.0865	0.0629	0.00200	2.768
^{192}Ir	0.127	0.128	0.268	0.266	0.0460	14.88
^{137}Cs	0.119	0.124	0.291	0.301	0.0426	14.28

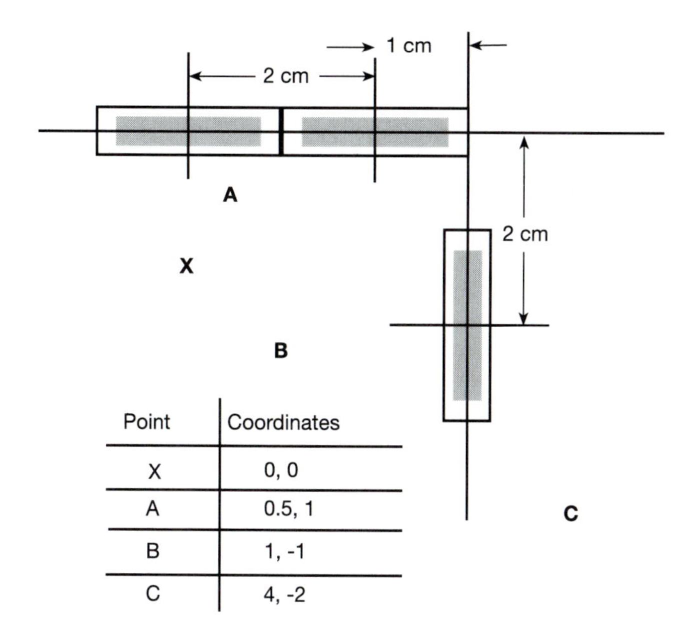

Figure 9.10. *A computation test case for a three-source configuration. Table 9.5 gives the dose rates for various types of source in this configuration.*

the computer's calculations to a standard, the numerical data should be used rather than dose distributions. Frequently the high gradients in a brachytherapy dose distribution obscure relatively large differences in actual values for the dose or dose rate.

Table 9.4 also contains values for the dose from a 5 d application. These values allow evaluation of the dose calculation algorithm that accounts for source decay *in situ*.

Table 9.5. *Dose rates for sources with the geometry shown in figure 9.10. All sources taken as 1 U. (Point source solutions include the anisotropy constant.)*

Source	Dose rates to points			
	X	A	B	C
^{137}Cs Amersham J	0.523	1.558	0.991	0.315
^{125}I model 6711	0.342	1.282	0.657	0.142
^{125}I model 6711 point source	0.329	1.207	0.627	0.169
^{103}Pd Theraseed	0.157	0.812	0.357	0.049
^{103}Pd Theraseed point source	0.147	0.762	0.351	0.071
^{192}Ir	0.551	1.583	0.918	0.305
^{192}Ir point source	0.540	1.561	0.885	0.311

As with the reconstruction algorithm, checks on the calculation algorithm only need performance during acceptance testing and following opening of the data files. However, checks on the calculational process should form part of the monthly evaluations. A stored set of 'standards' consisting of a plan of each general type commonly encountered at the facility (such as prostate implants, eye plaques etc) should be stored, recalled and recalculated each month to provide an assurance that the calculation formulae and factors remain unchanged. The standard selected can be rotated through a cycle to cover the range as long as the cycle does not exceed six months. If the computer offers a checksum facility for the data files, a monthly check compared with the value at the time on installation provides some assurance that the files remain the same.

One problem in some treatment planning systems arises following rotations of the source set (or point of view, whichever moves). Due to rounding of transformed coordinates, multiple rotations may result in deviations of the source (or target) coordinates. To test for this problem, enter a set of source coordinates. Several sources in a simple geometric configuration works well. Rotate this arrangement several times, but avoiding rotations through near 90 and 180 degrees, eventually coming back to the original orientation. Print the coordinates of the sources and compare with those entered. Differences greater than 0.2 mm in any coordinate may indicate that successive rotations may produce shifts in the source position.

An increasing number of systems include optimization routines, and these routines optimize the source distributions from an increasing number of approaches. While critical to the treatment planning processes, few of the optimization approaches lend themselves to manual verification or review. Fortunately, the processes need not undergo manual checking, since it is the final product of the process that is of interest, and this should be evaluated during each per patient review. However, part of the monthly tests run during the recall and recalculation of plans described two paragraphs earlier should include a reoptimization to assure that the algorithm continues to perform in an unchanged manner. During the training period for a treatment planning system with optimization capabilities, each operator needs to become familiar with the effects of varying all input parameters for the optimization, and what the algorithm produces when faced with widely varying input source geometries.

Very few treatment-planning computers include correction of dose distributions for shields and other inhomogeneities, and even fewer make these corrections correctly. If the computation algorithm includes such corrections, and the user intends to apply them to clinical cases, the effects of the corrections require establishing. Unfortunately, such testing usually requires phantoms with inhomogeneities and a method for measurement in the phantom, and, as mentioned previously, measurement of brachytherapy doses poses considerable challenges. Hand calculations for inhomogeneities usually prove as inadequate as most computer algorithms. With no good method to evaluate the computer calculations and no reason to believe the calculations correct, the user is left with taking such corrected calculations with a large grain of salt. That the doses cannot be calculated does not mean that shields should not be used; in many cases the shield probably produces dose distributions closer to that desired than omitting them. The lack of practical quality management on the calculation simply reflects the state of the art.

9.2.4. Plan output

The output from most treatment planning systems consists of dose distributions and data sheets containing information about the plan. Even when both functions utilize the same piece of equipment, they often use different driving software, and need separate consideration.

9.2.4.1. Plots. Plotters may have two problems:

- *Alignment of coordinate systems.* A plot may use a different file to describe the lines for the dose distribution and the source coordinates. Patient anatomical points may come from an entirely different file. The coordinates for each of the files must be coordinated both in position and rotation. This particularly becomes challenging if the various pieces of information come from different imaging inputs. This is best tested using a phantom with several radiopaque markers in known locations, imaged as customary for source and patient localization. Assigning some of the markers as sources and others as patient anatomy allows calculation of the relative positions of all points, and coordination of the various information files.
- *Scaling.* The scale on the plot may be different from that written by the plotter, or that specified when sending the dose distribution to the plotter. The latter proves to be an irritation, the former a hazard. To evaluate the scaling of the dose distribution, enter a single source of a type used in the

Table 9.6. *Information a brachytherapy treatment planning computer system should contain.*

Patient's name and identification number
Date of the calculation and of data entry if different
Prescribed dose or dose rate
Source locations if free sources
Needle or catheter locations if sources confined, with source position in catheter or needle
Optimization technique and factors
Dose rates and/or doses to specified points and the coordinates of those points
Values used in the calculation algorithm
Type of reconstruction and the basic data for calculating coordinates

For low dose rate applications
- Treatment duration
- Strength of sources

For high dose rate applications
- Strength of source
- Dwell times
- Treatment unit settings
  - Channel number
  - Length
  - Step size

test of the computation algorithm, with its axis in the display plane. Instruct the computer to plot the dose rate line corresponding to the value at 5 cm. In the plot, this line surrounds the source, and its intersection of the source perpendicular bisector on the two sides should be 10 cm apart to within a few tenths of a millimetre. Problems with scaling frequently occur with postscript colour plotters with automatic rescaling to fit the page.

As the main source of dose distribution information, the accuracy of the plotter should be evaluated monthly, even though several of the checks performed for individual patient plans should alert the physics staff to problems with the plotter. To assess changes in the plotter, recall a standard dose distribution, generate a plot and compare the plot to one generated during the acceptance testing of the plotter.

9.2.4.2. Printed output. The printed output should contain all of the information necessary for evaluation of correctness of the plan and then execution. Table 9.6 lists the items the plan should contain. Curran and Starkschall (1991) recommend evaluating the printer for print quality monthly.

9.3. SUMMARY

Each of the many steps required to complete a dose calculation presents opportunities for errors to enter the treatment plan. Quality management for dose calculations in brachytherapy addresses the two phases of the process: localization (and reconstruction) and dose calculation.

Brachytherapy localization requires first-rate imaging equipment. Each type of imaging presents its own characteristics and requirements for quality management. Radiographic localization requires acceptance and periodic testing of the coincidence of the cross hairs with the designated axis and angular accuracy. In each case, a check should be built into the system to evaluate the critical parameters for the type of reconstruction used. The sections above discuss various types of consistency monitor possible.

For sectional imaging (US, CT, MRI) acceptance and periodic testing should evaluate correct indication of image planes, image distortion and values for coordinates passed to the planning computer.

Correct localization requires accurate simulation of source positions. To achieve this, the markers must correctly mimic the sources in any important way, particularly with respect to position in the appliance.

The dose calculation engine requires considerable testing upon commission of a treatment planning system, not only to evaluate the accuracy with which the system performs the calculation, but so the operator becomes familiar with all the characteristics of the algorithms and program. Test cases and manual checks on calculations in simple cases provide initial assessments of the system. After verifying that the system operates correctly, saving the output for some standard plans provides benchmarks for future comparisons.

Acceptance and periodic testing also must test the accuracy and adequacy of all relevant peripheral devices, including the digitizers, printers and plotters.

APPENDIX 9A

The following are tables of the anisotropy functions for some selected brachytherapy sources. These tables from AAPM TG43 appear with the permission of the American Association of Physicists in Medicine.

Table 9.7. *Anisotropy function, $F(r, \theta)$, for a ^{192}Ir source (stainless steel encapsulation).*

r (cm)	$\theta = 0.0°$	$\theta = 10.0°$	$\theta = 20.0°$	$\theta = 30.0°$	$\theta = 40.0°$
1.0	0.806	0.843	0.947	0.966	1.000
2.0	0.788	0.906	0.947	0.941	0.945
3.0	0.769	0.813	0.893	0.936	1.030
4.0	0.868	0.949	1.010	0.996	1.020
5.0	0.831	0.931	0.994	1.030	1.070
6.0	0.819	0.899	0.920	0.928	0.973
7.0	0.844	0.944	0.985	0.969	0.962

r (cm)	$\theta = 50.0°$	$\theta = 60.0°$	$\theta = 70.0°$	$\theta = 80.0°$	$\theta = 90.0°$
1.0	1.020	1.030	1.030	1.020	1.000
2.0	0.949	0.953	0.989	0.991	1.000
3.0	0.984	0.977	1.030	1.030	1.000
4.0	1.03	1.040	1.010	1.010	1.000
5.0	1.05	1.030	1.010	1.020	1.000
6.0	0.959	0.954	0.996	0.997	1.000
7.0	0.967	1.010	1.020	0.979	1.000

Table 9.8. *Anisotropy function, $F(r, \theta)$, for a ^{125}I model 6711 source.*

r (cm)	$\theta = 0.0°$	$\theta = 10.0°$	$\theta = 20.0°$	$\theta = 30.0°$	$\theta = 40.0°$
1.0	0.350	0.423	0.628	0.826	0.953
2.0	0.439	0.549	0.690	0.816	0.903
3.0	0.452	0.529	0.612	0.754	0.888
4.0	0.521	0.582	0.688	0.798	0.842
5.0	0.573	0.600	0.681	0.793	0.888
6.0	0.581	0.621	0.718	0.803	0.826
7.0	0.603	0.660	0.758	0.829	0.861

r (cm)	$\theta = 50.0°$	$\theta = 60.0°$	$\theta = 70.0°$	$\theta = 80.0°$	$\theta = 9.0°$
1.0	1.00	1.03	1.04	1.02	1.00
2.0	0.954	1.00	1.06	1.04	1.00
3.0	0.795	0.877	1.07	1.07	1.00
4.0	0.866	0.935	0.969	0.998	1.00
5.0	0.903	0.922	0.915	0.985	1.00
6.0	0.891	0.912	0.953	0.962	1.00
7.0	0.922	0.932	0.978	0.972	1.00

Table 9.9. *Anisotropy function, $F(r, \theta)$, for a ^{103}Pd model 200 source.*

r (cm)	$\theta = 0.0°$	$\theta = 10.0°$	$\theta = 20.0°$	$\theta = 30.0°$	$\theta = 40.0°$
1.0	0.523	0.575	0.700	0.741	0.843
2.0	0.526	0.544	0.590	0.615	0.748
3.0	0.533	0.575	0.520	0.572	0.791
4.0	0.544	0.442	0.463	0.646	0.773
5.0	0.624	0.574	0.627	0.758	0.777
r (cm)	$\theta = 50.0°$	$\theta = 60.0°$	$\theta = 70.0°$	$\theta = 80.0°$	$\theta = 90.0°$
1.0	0.939	1.000	1.030	1.020	1.000
2.0	0.876	0.962	1.080	1.060	1.000
3.0	0.658	0.838	1.040	0.948	1.000
4.0	0.848	0.887	0.941	0.963	1.000
5.0	0.901	0.939	1.040	0.992	1.000

Reproduced from AAPM TG43 (1995).

REFERENCES

Abelquist E W 1998 Use of smears for assessing removable contamination *Health Phys. Newslett.* **26** 18–19

Altshuler B and Pasternack B 1963 Statistical measures of the lower limit of detection of a radioactive counter *Health Phys.* **9** 293–8

American Association of Physicists in Medicine: Task Group 40 (AAPM TG40) (Kutcher G, Coia L, Gillin M, Hanson W, Leibel S, Morton R, Palta J, Purdy J, Reistein L, Svensson G, Weller M and Wingfield L) 1994 Report 46: Comprehensive QA for radiation oncology *Med. Phys.* **21** 581–618

American Association of Physicists in Medicine: Task Group 43 (AAPM TG43) (Nath R, Anderson L L, Luxton G, Meli J A, Weaver K, Williamson J F and Meigooni A) 1995 Dosimetry of interstitial brachytherapy sources *Med. Phys.* **22** 209–34

American Association of Physicists in Medicine: Task Group 53 (AAPM TG53) (Fraass B, Doppke K, Hunt M, Kutcher G, Starkschall G, Stern R and van Dyke J) 1998 Quality assurance for clinical radiotherapy treatment planning *Med. Phys.* **25** 1773–1829

American Association of Physicists in Medicine: Task Group 56 (AAPM TG56) (Nath R, Anderson L L, Meli J A, Olch A J, Stitt J A and Williamson J F) 1997 Code of practice for brachytherapy physics *Med. Phys.* **24** 1557–98

American Association of Physicists in Medicine: Task Group 59 (AAPM TG59) (Kubo H D, Glasgow G P, Pethel T D, Thomadsen B R and Williamson J F) 1998 High dose-rate brachytherapy delivery *Med. Phys.* **25** 375–403

American Society for Quality Control (ASQC) 1998 Web site 17 November (http://www.asqc.org)

Anderson L L 1976 Spacing nomograph for interstitial implants of ^{125}I seeds *Med. Phys.* **3** 48–51

——1986 A 'natural' volume–dose histogram for brachytherapy *Med. Phys.* **13** 899–903

Anderson L L, Hilaris B S and Wagner L K 1985 A nomograph for planar implant planning *Endocuriether./Hypertherm. Oncol.* **1** 9–15

Anderson L L, Nath K, Weaver K A, *et al* 1990 *Interstitial Brachytherapy: Physical, Biological and Clinical Considerations* (New York: Raven)

Anderson L L, Moni J V and Harrison L B 1993 A nomography for permanent implants of ^{103}Pd seeds *Int. J. Radiat. Oncol. Biol. Phys.* **27** 129–35

Bastin K T, Podgorsak M S and Thomadsen B R 1993 The transit dose component of high dose-rate brachytherapy: direct measurements and clinical implications *Int. J. Radiat. Oncol. Biol. Phys.* **26** 695–702

Bernard M and Dutreix A 1968 Dispostif de mesure de l'activité lineique *J. Radiol. Electrol. Med. Nucl.* **49** 534–8

Campbell J L, Santerre C R, Farina P C and Muse L A 1993 Wipe testing for surface contamination by tritiated compounds *Health Phys.* **64** 540–4

Chenery S G A, Pla M and Podgorsak E B 1985 Physical characteristics of the Selectron high dose rate intracavitary afterloader *Br. J. Radiol.* **58** 735–40

College of American Pathologists (CAP) 1987 Appendix O *Standards for Laboratory Accreditation* (Northfield, IL: College of American Pathologists)

Curran B and Starkschall G 1991 A program for quality assurance of dose planning computers *Quality Assurance in Radiotherapy Physics* ed G Starkschall and J Horton (Madison, WI: Medical Physics) pp 207–28

Currie L A 1968 Limits for qualitative detection and quantitative determination: application to radiochemistry *Anal. Chem.* **40** 586–93

DeWerd L A 1995 Source strength standards and calibration of HDR/PDR sources *Brachytheraphy Physics* ed J F Williamson, B R Thomadsen and R Nath (Madison, WI: Medical Physics) pp 541–55

DeWerd L A, Jursinic P, Kitchen R and Thomadsen B R 1995 Quality assurance tool for high dose rate brachytherapy *Med. Phys.* **22** 435–40

Evens M D C, Arsenault C J, Cygler J and Laewen A 1993 Quality assurance for variable-length catheters with an afterloading brachytherapy device *Med. Phys.* **20** 251–4

Ezzell G A 1990 Acceptance testing and quality assurance for high dose-rate remote afterloading systems *Brachytherapy HDR and LDR* ed A A Martinez, C G Orton and R F Mould (Columbia, MD: Nucletron) pp 138–59

——1991 Quality assurance in HDR brachytherapy: physical and technical aspects *Selectron Brachyther. J.* **5** 59–62

——1994 Quality assurance of treatment plans for optimized high dose rate brachytherapy *Med. Phys.* **21** 659–61

——1999 Gynecological application III: HDR approaches to cancer of the cervix *American Brachytherapy Society School of Brachytherapy Physics (Gainsville, FL, 1999)*

Ezzell G A and Luthmann R W 1995 Clinical implementation of dwell-time optimization techniques for single stepping-source remote afterloaders *Brachytherapy Physics* ed J F Williamson, B R Thomadsen and R Nath (Madison, WI: Medical Physics) pp 617–40

Feder B H, Syed A M N and Neblett D 1978 Treatment of extensive carcinoma of the cervix with the 'transperineal parametrial butterfly' *Int. J. Radiat. Oncol. Biol. Phys.* **4** 735–42

Fitzgerald L T and Mauderli W 1975 Analysis of errors in three-dimensional reconstruction of radium implants from stereo radiographs *Radiology* **115** 445–58

Fleming P, Buchler D, Higgins P, Bak M and Thomadsen B 1984 Cumulative radiation effect (CRE) as a potential guide in the management of advanced carcinoma of the vulva *Optimization of Cancer Radiotherapy* ed B R Paliwal, D E Herbert and C G Orton, pp 65–82

Fleming P, Syed A M N, Neblett D, Puthawala A, George F W III and Townsend D 1980 Description of an afterloading ^{192}Ir interstitial–intracavitary technique in the treatment of carcinoma of the vagina *Obstet. Gynecol.* **55** 525–30

Flynn A 1990 Quality assurance checks on a MicroSelectron-NDR *Selectron Brachyther. J.* **4** 112–15

Glasgow G P 1995 Principles of remote afterloading devices *Brachytherapy Physics* ed J F Williamson, B R Thomadsen and R Nath (Madison, WI: Medical Physics) pp 485–502

Godden T 1988 *Physical Aspects of Brachytherapy* (Bristol: Institute of Physics Publishing)

Goetsch S J, Attix F H, DeWerd L A and Thomadsen B R 1992 A new re-entrant ionization chamber for the calibration of iridium-192 high-dose-rate sources *Int. J. Radiat. Oncol. Biol. Phys.* **24** 167–70

Goetsch S J, Attix F H, Pearson D W and Thomadsen B R 1991 Calibration of high-dose-rate afterloading systems *Med. Phys.* **18** 462–7

Grigsby P W 1989 Quality assurance of remote afterloading equipment at the Mallinckrodt Institute of Radiology *Selectron Brachyther. J.* **1** 15

Gryna F M 1988 Quality assurance *Juran's Quality Control Handbook* 4th edn, ed J M Juran and F M Gryna (New York: McGraw-Hill) p 9.2

Haas J S, Dean R D and Mansfield C M 1985 Dosimetric comparison of the Fletcher family of gynecologic colpostats 1950–1980 *Int. J. Radiat. Oncol. Biol. Phys.* **11** 1317–21

Hicks J A and Ezzell G A 1995 Calibration and quality assurance *Activity, Special Report 7: Quality Assurance* (Veenendaal: Nucletron–Oldelft) pp 15–24

Hunter R D 1995 The Manchester experience with LDR variation in brachytherapy of cervix carcinoma *International Brachytherapy: 8th Int. Brachytherapy Conf. (Nice, 1995)* (Veenendaal: Nucletron–Oldelft) pp 52–5

International Commission on Radiation Units (ICRU) 1993 Prescribing, recording, and reporting photon beam therapy *ICRU Report* 50

——1997 Dose and volume specification for reporting interstitial therapy *ICRU Report* 58

ISO 8402:1994 quoted by Peach R W 1992 *The ISO 9000 Handbook* (New York: McGraw-Hill)

Jones C H 1990 Quality assurance in brachytherapy using the Selectron LDR/MDR and MicroSelectron-HDR *Selectron Brachyther. J.* **4** 48–52

Juran J M 1988 The quality function *Juran's Quality Control Handbook* 4th edn, ed J M Juran and F M Gryna (New York: McGraw-Hill) pp 2.1–2.13

King C C, Stockstill T F, Bloomer W D, Kalnicki S, Wu A, Buchsbaum R and Chen A S 1992 Point dose variations with time in brachytherapy for cervical carcinoma *Med. Phys.* **19** 777

Klein R C I, Linnins E L and Gershey 1992 Detecting removable surface contamination *Health Phys.* **62** 186–9

Kubo H 1992 Verification of treatment plans by mathematical formulas for single catheter HDR brachytherapy *Med. Dosim.* **17** 151–5

Kubo H and Chin R B 1992 Simple mathematical formulas for quick-checking of single-catheter high dose rate brachytherapy treatment plans *Endocuriether./Hypertherm. Oncol.* **8** 165–9

Kwan D K, Kagan A R, Olch A J, Chan P Y, Hintz B L and Wollin M 1983 Single- and double-plane iridium-192 interstitial implants: implantation guidelines and dosimetry *Med. Phys.* **10** 456–61

Ling C 1995 Radiobiological considerations in brachytherapy *Brachytherapy Physics* ed J F Williamson, B R Thomadsen and R Nath (Madison, WI: Medical Physics) pp 39–69

Meigooni A S, Williamson J F and Slessinger E D 1992 Practical quality assurance tests for positional and temporal accuracy of HDR remote afterloaders (abstract) *Endocuriether./Hypertherm. Oncol.*

Mellenberg D E Jr and Kline R W 1995 Verification of manufacturer-supplied ^{125}I and ^{103}Pd air-kerma strengths *Med. Phys.* **22** 1495–7

Mould R F, Battermann J J, Martinez A A and Speiser B L 1994 *Brachytherapy from Radium to Optimization* (Veenendaal: Nucletron)

Nag S 1994 *High Dose Rate Brachytherapy: a Textbook* (Armonk, NY: Futura)

——1997 *Principles and Practice of Brachytherapy* (Armonk, NY: Futura)

National Council on Radiation Protection and Measurement (NCRP) 1978 *Report 58: a Handbook of Radioactivity Measurements Procedures* (Washington, DC: NCRP)

National Council on Radiation Protection and Measurement (NCRP) 1988 *Report 99: Quality Assurance for Diagnostic Imaging* (Washington, DC: NCRP)

Neblett D, Clinical techniques and applicators for interstitial implantation *Brachytherapy Physics* ed J F Williamson, B R Thomadsen and R Nath (Madison, WI: Medical Physics) pp 281–300

Neblett D L, Syed A M N, Puthawala A A, Harrop R, Frey H S and Hogan S E 1985 An interstitial implant technique evaluated by contiguous volume analysis *Endocuriether./Hypertherm. Oncol.* **1** 213–21

Niroomand-Rad A and Thomadsen B R 1990 Evaluation of the reconstruction of seed position from stereo and orthogonal radiography for routine radiotherapy planning *Radiat. Med.* **8** 145–51

Niroomand-Rad A, Thomadsen B R and Vainio P 1987 Evaluation of the reconstruction of brachytherapy implants in three-dimensions from stereo radiographys *Radiother. Oncol.* **8** 337–42

Noyes W R, Bastin K T, Edwards S A, Buchler D A, Stitt J A, Thomadsen B R, Fowler J F and Kinsella T J 1995a Postoperative vaginal cuff irradiation using high dose rate remote afterloading: a phase II clinical protocol *Int. J. Radiat. Oncol. Biol. Phys.* **32** 1439–43

Noyes W R, Peters N E, Thomadsen B R, Fowler J F, Buchler D A, Stitt J A and Kinsella T J 1995b Impact of 'optimized' treatment planning for tandem and ring, and tandem and ovoids, using high dose rate brachytherapy for cervical cancer *Int. J. Radiat. Oncol. Biol. Phys.* **31** 79–86

Paliwal B R, Thomadsen B R and Petereit D G 1997 Imaging applications in brachytherapy *Principles and Practice of Brachytherapy* ed S Nag (Armonk, NY: Futura) pp 201–17

Pirsig R M 1974 *Zen and the Art of Motorcycle Maintenance* (New York: Morrow)

Podgorsak M B, Paliwal B R, Thomadsen B R, Stitt J A and Buchler D A 1993 Radiographic visualization of vaginal cylinders in gynecologic high dose rate brachytherapy *Int. J. Radiat. Oncol. Biol. Phys.* **25** 525–7

Potash R A, Gerbi B J and Engeler G P 1995 *Brachytherapy Physics* ed J F Williamson, B R Thomadsen and R Nath (Madison, WI: Medical Physics) pp 379–410

Ritter M, Shahabi S, Gehring M, Shanahan T, Thomadsen B and Kinsella T 1989 Transperineal prostate implantation with three-dimensional, computed tomography-based preplanning and customized template design *American Endocurietherapy Society (Hilton Head, SC, 1989)*

Rogus R D, Smith M J and Kubo H D 1998 An equation to QA check the total treatment time for single-catheter HDR brachytherapy *Int. J. Radiat. Oncol. Biol. Phys.* **40** 245–8

Saw C B and Suntharalingam N 1991 Qualitative assessment of interstitial implants *Int. J. Radiat. Oncol. Biol. Phys.* **20** 135–9

Saylor W and Dillard M 1976 Dosimetry of cesium-137 sources with the Fletcher–Suit gynecologic applicator *Med. Phys.* **3** 117–19

Shalek R and Stoval M 1969 Dosimetry in implant therapy *Radiation Dosimetry* 2nd edn, vol VIII, ed F H Attix and E Tochilin (New York: Academic) pp 745–68

Sharma S C, Williamson J F and Cytacki E 1982 Dosimetric analysis of stereo and orthogonal reconstruction of interstitial implants *Int. J. Radiat. Oncol. Biol. Phys.* **8** 1803–5

Slessinger E D 1989 Calibration of differential linear activity cesium sources (abstract) *Med. Phys.* **16** 495

Slessinger E D 1990a A quality assurance program for low dose rate remote afterloading devices *Brachytherapy HDR and LDR* ed A A Martinez, C G Orton and R F Mould (Columbia, MD: Nucletron) pp 160–8

——1990b Selectron-LDR quality assurance *Activity* **5** 17–21

——1995a Commissioning of non-stepping source remote afterloaders *Brachytherapy Physics* ed J F Williamson, B R Thomadsen and R Nath (Madison, WI: Medical Physics) pp 503–22

——1995b Clinical implementation of LDR remote afterloading *Brachytherapy Physics* ed J F Williamson, B R Thomadsen and R Nath (Madison, WI: Medical Physics) pp 521–40

Speiser B L and Hicks J A 1995 Safety programmes for remote afterloading brachytherapy: high dose rate and pulsed low dose rate *Activity, Special Report 7: Quality Assurance* (Veenendaal: Nucletron–Oldelft) pp 3–11

Stitt J A, Fowler J F, Thomadsen B R, Buchler D A, Paliwal B R, Shahabi S and Kinsella T J 1992 High dose-rate intracavitary brachytherapy for carcinoma of the cervix: the Madison system: I. Clinical and radiobiological considerations *Int. J. Radiat. Oncol. Biol. Phys.* **24** 335–48

Thomadsen B R 1995a Clinical implementation of HDR intracavitary and transluminal brachytherapy *Brachytherapy Physics* ed J F Williamson, B R Thomadsen and R Nath (Madison, WI: Medical Physics) pp 641–78

——1995b Clinical implementation of remote-afterloading brachytherapy *Brachytherapy Physics* ed J F Williamson, B R Thomadsen and R Nath (Madison, WI: Medical Physics) pp 680–98

Thomadsen B R, Ayyanger K, Anderson L L, Luxton G, Hanson W and Wison J F Brachytherapy treatment planning *Principles and Practice of Brachytherapy* ed S Nag (Armonk, NY: Futura) pp 127–99

Thomadsen B R, Caldwell B, Stitt J A, McConley R and Leammerich P 1998 Toward understanding human factors in medicine *Trans. Am. Nucl. Soc.* **79** 76–9

Thomadsen B R, DeWerd L A and McNutt T 1997 Assessment of the strength of individual ^{192}Ir seeds in ribbons *Med. Phys.* **24** 1019

Thomadsen B R and Hendee E G 1999 Brachytherapy radionuclides, dosimetry and dose distributions *Biomedical Uses of Radiation* ed W Hendee (Weinheim: Wiley–VCH)

Thomadsen B R, Hudek P V, van der Laarse E, Kolkman-Deurloo I-KK and Visser A G 1994 Treatment planning and optimization *Textbook of High Dose-Rate Brachytherapy* ed S Nag (Mt Kisco, NY: Futura)

Thomadsen B R, Shahabi S, Stitt J A, Buchler D A, Fowler J F, Paliwal B R and Kinsella T J 1992 High dose-rate intracavitary brachytherapy for carcinoma of the cervix: the

Madison system: II. Procedural and physical considerations *Int. J. Radiat. Oncol. Biol. Phys.* **24** 349–57

United States Department of Health, Education and Welfare (USDHEW) Public Health Service 1970 *Radiological Health Handbook* (Washington, DC: US Government Printing Office)

Williams J R and Thwaites D L 1993 *Radiotherapy Physics in Practice* (Oxford: Oxford University Press)

Williamson J F 1991 Practical quality assurance for low dose-rate brachytherapy *Quality Assurance in Radiotherapy Physics* ed G Starkschall and J Horton (Madison, WI: Medical Physics) pp 139–82

Williamson J F 1995 Clinical physics of pulsed dose-rate remotely afterloaded brachytherapy *Brachytherapy Physics* ed J F Williamson, B R Thomadsen and R Nath (Madison, WI: Medical Physics) pp 577–616

——1998a Monte Carlo-based dose-rate tables for the Amersham CDCS-J and 3M model 6500 ^{137}Cs tubes *Int. J. Radiat. Oncol. Biol. Phys.* **41** 959–70

——1998b Physics of brachytherapy *Principles and Practice of Radiation Oncology* 3rd edn, ed C Perez and L Brady (Philadelphia, PA: Lippincott–Raven) pp 405–67

Williamson J F, Ezzell G A, Olch A and Thomadsen B R 1994 Quality assurance for high dose-rate brachytherapy *Textbook of High Dose-Rate Brachytherapy* ed S Nag (Mt Kisco, NY: Futura)

Williamson J F, Thomadsen B R and Nath R 1995 *Brachytherapy Physics* (Madison, WI: Medical Physics)

Zwicker R D and Schmidt-Ullrich R 1995 Dose uniformity in a planar interstitial implant *Int. J. Radiat. Oncol. Biol. Phys.* **31** 149–55

Zwicker R D, Schmidt-Ullrich R and Schiller B 1985 Planning of Ir-192 seed implants for boost irradiation to the breast *Int. J. Radiat. Oncol. Biol. Phys.* **11** 2163–70

INDEX